Newnes

Audio & Hi-fi Engineer's

Pocket Book

Other books by Vivian Capel

Home Security
Security Systems and Intruder Alarms
Public Address Systems
An Introduction to Loudspeaker and Enclosure Design
Loudspeakers for Musicians
Public Address Loudspeaker Systems
Acoustic Feedback — How to Avoid It

Newnes

Audio & Hi-fi Engineer's Pocket Book

3rd edition

Vivian Capel

Newnes
An imprint of Butterworth-Heinemann Ltd
Linacre House, Jordan Hill, Oxford OX2 8DP

A member of the Reed Elsevier group

OXFORD   LONDON   BOSTON
MUNICH   NEW DELHI   SINGAPORE   SYDNEY
TOKYO   TORONTO   WELLINGTON

First published 1988
Reprinted 1988
Second edition 1991
Reprinted 1993
Third edition 1994

British Library Cataloguing in Publication Data
A catalogue record for this book is available from the
British Library

ISBN 0 7506 2001 3

Library of Congress Cataloguing in Publication Data
A catalogue record for this book is available from the
Library of Congress

Produced by Co-publications, Loughborough
— part of The Sylvester Press
Printed in Great Britain by Clays Ltd, St Ives plc

Contents

Human hearing: the ear; logarithmic frequency and
amplitude response; equal loudness contours; hearing
damage; presbycusis. Sound sources: pure tones; harmonics;
nodes and antinodes; musical instruments-frequency ranges,
harmonic content. Measuring sound: the decibel; sound
intensity; sound pressure; loudness; the phon; pressure levels
of common sounds; weighting curves; particle velocity;
acoustic impedance; sound power level. Sound propagation:
point source; dipole; line source; plane source; directivity
factor; speed of sound; wavelength/frequency; near/far
fields; reverberant field; room radius; reflection; diffraction;
refraction; effects of wind and humidity; energy
transmission; atmospheric absorption. Room and hall
acoustics; reverberation time; speech clarity test; absorption
coefficients; absorbers; assisted reverberation; resonances;
golden ratio; flutter and echo; reflectors. Sound insulation:
source propagation; wall insulation; sound reduction
characteristics; damping; mass law; cavity walls; flanking
walls; ceilings and floors; doors and windows; ventilation
systems.

Transducers: moving-coil; ribbon; ceramic; capacitor; r.f.
capacitor; electret; phantom powering; A-B powering.
Acoustic characteristics: omni; cardioid; hypercardioid;
velocity; interference tube; directivity coefficient; sound
power concentration; ambient noise. Special types: multi-
pattern, zoom, noise cancelling; parabolic reflectors.
Boundary microphones. Microphones in the sound field;
attenuation with distance; proximity effects; pressure
buildup; direct/diffuse fields; multiple sources; stereo;
popping. Electrical parameters: Sensitivity; impedance;
cable capacitance; frequency resonse; electrical noise; hum;
distortion; input circuits; applications; connectors.

Tape transport: main drive; rewind; take-up; dual capstan drive; reversing systems. Motors: speed control; governors; ac control; brushless motors; speed adjustment. Hi-Fi from Video: f.m. recording; Beta Hi-fi; VHS Hi-Fi. Digital tape recording: analogue to digital conversion; processors; digital tape; digital formats. Philips digital cassettte: DCC head; PASC encoding. Recording tape: base; coatings; particles; chromium dioxide; metal tape; thin-film coatings. Tape manufacture. Musicassette manufacture. Tape parameters. Tape groups. Open reel recording. Compact cassette.

Receiver principles; frequency modulation; bandwidth; i.f.stages; ceramic filters; pre-emphasis; demodulators; phase-locked loop. Stereo broadcasting: L and R signals; subcarrier; pilot tone; modulation spectrum. Decoding: subcarrier regeneration; synchronous detection. BBC stereo tests; noise. Interference. Alignment. Tuner parameters. F.M. aerials: dipoles; impedance; voltage/current distribution; folded dipole; directivity; cable impedance.

Valves and their features. Valve circuits. Servicing valve amplifiers; disturbance testing. Vintage radio: pre-war valves; post-war. Audio transistors: point contact; junction; alloy-diffused; planar; epitaxial; fet; mos fet; vmos fet. Transistor characteristics: alpha; beta; bias; heat dissipation; H_{FE}/H_{fe}; limiting parameters. Basic circuit features; common base; common emitter; common collector; class A; class B; variants; distortion causes; class C; class D; pulse-width modulation; darlington pair; differential pair; driver stage; feedforward; input stages; matching; negative feedback; overloading; h parameters; resistor noise; threshold voltage; tone controls; equalisers; gyrators. Amplifier parameters.

The driver; doppler effect; phasing; controlled flexure; cone materials; suspension; cone resonance; damping; radiation resistance; delayed resonance; bell mode; laser interferometer; speech coil; tweeter cones and domes. Enclosures. Doublet. Infinite baffle; air resonances; panel resonances; reflex; auxiliary bass radiators. Horn; flares. Labyrinth; transmission line; resonant pipe. Line source; response attenuation. Absorbents; isothermal propagation. Electrostatic speakers: bias; charge migration; polarising voltage, frequency response. Electronic instrument

loudspeakers: guitar, keyboard, monitor, vocal, cabaret.
Unconventional speakers: orthodynamic; air motion
transformer; ribbon tweeter; piezo tweeter; plasma tweeter.
Crossover networks: filter order; crossover frequencies;
circuits; disadvantages. The full-range single-driver.

Direction location: aural time delays; amplitude difference;
head masking; stereo recording: coincident and spaced
microphones, pan-potting; ambience, Haffler effect. Quad
systems: SQ, QS, CD-4. NRDC Ambisonic system; Matrix
H. Dolby stereo. Surround-sound systems: Zuccarelli
holophony; Sensaura; Roland Sound; stereo image
extension.

Objectives. Loudspeaker distribution; 100-volt lines;
transformer tappings. Avoiding feedback; choice of
microphones; absorption; loudspeaker installation;
frequency shifter. Delay lines, digital and analogue; bucket-
brigade devices; clock frequency. Uses; Haas effect. Ceiling
speakers; mixer features. LISCA loudspeaker system.
Mixers. Open-air classical concerts: time delay;
loudspeakers; microphones; stage canopy design.
Microphones. Induction loops; current; resistance; height;
power. System problems; instability; earth loops; radio
interference; cable damage. Safety regulations.

Resistance, reactance, power and impedance; Q factor; skin
effect; resonance; time constants. Attenuators L-type;
T-type; π-type; symmetrical and asymmetrical. Filters:
constant-k; m-derived, high and low pass. Test equipment:
sound, mV, output, distortion meters; audio, pink noise
generators. Oscilloscope. Power circuits; transistor
connections.

Preface to the third edition

When the second edition went to press just over two years ago it seemed that there was little scope for a further edition for some considerable time. Yet since then there have been a number of developments and areas of growing interest in the audio field which need to be covered.

The Sony Mini Disc has arrived employing magneto-optical technology that is totally new to audio, although used in computer data storage. Information on this has therefore been added to the CD section.

Surround sound is again in the news with the Sensaura system that is claimed to give a sensation of sound from all directions, using only two stereo channels, without any need of decoders for replay. If it is anywhere near as good as its predecessor, Zuccarelli's holophony, on which it appears to be based, it could be the major audio event of the decade.

Another surround system the Roland Sound Space has also surfaced; it uses electronic rather than the acoustic encoding of Sensaura. To acquaint readers with the principles of sound-source location, with a look at previous attempts to achieve it, including the ill-fated quad systems, the NRDC Ambisonic system, and Zuccarelli's holophony as well as the present ones, a complete new section on Surround Sound has been added.

Valve amplifiers continue to be popular among the enthusiasts, so the section on valves has been expanded to include amplifier circuit features and hints on servicing. Vintage radio is another area of growing interest; so a section has been included outlining the principles of pre-war and post-war valve radio receivers with servicing tips.

Another growing phenomena is the outdoor classical music concert, for which sound engineers may be required to provide sound reinforcement. The requirements are quite different from the pop music concert covered in the second edition. Basic principles are given to help the engineer for whom this is new ground.

The information previously given on crossover networks has been balanced by a discussion of some of their little-known evils and those of multi-driver loudspeakers, in a consideration of the virtues of the single full-range driver. This should provide some food for thought.

Finally, the new safety regulations for public-address systems are dealt with, and how they affect installation and general practice. Some are to be welcomed as improving safety, but others are nonsensical and will prove an unnecessary burden to installers. Unfortunately all have to be observed.

With many new editions, material from previous editions is abandoned to make room for the fresh. This has not been done here, so it is hoped that this edition will give an even wider coverage of the audio scene than the previous ones, and so be of value to all in the industry as well as students and non-professional readers who just want to be informed on all things audio.

Vivian Capel

Sound and acoustics

Human hearing

The final link in the sound reproduction chain is the human ear. A basic knowledge of how it works can help in the understanding of many of the problems encountered in the field of audio engineering.

Directivity Sounds arriving from a source situated to one side of a listener arrive at the furthest ear fractionally later than at the nearest. There is thus a delay in phase which the brain interprets in terms of direction. With long waves (low frequencies) the phase difference is slight, but is greater with short-wave high frequencies. This is why it is difficult to locate a low-frequency source while a high-frequency one can be pinpointed with ease.

The convolutions of the *pinna* or outer ear produce reflections and phase delays which differ according to the angle of incidence of the sound wave. This also aids in direction location, especially it is believed, adding information as to the vertical position. Again only higher frequencies are effectively so identified.

At mid-frequencies, the masking effect of the head plays a part by introducing small amplitude differences.

Impedance matching; amplitude control The ear drum or *tympanum* vibrates in sympathy with the received sound and transmits the vibrations through three small bones called the *ossicles* in the middle ear. These are the *hammer*, *anvil* and *stirrup* or *stapes* as it is also called. The first two form a pair of pivoted levers that produces a nominal leverage ratio of 3:1, the third communicates the vibrations to the window of the inner ear. This ratio matches the mechanical impedance of the ear drum to that of the window, so obtaining optimum power transfer.

The tiny muscles holding the bones in place permit the pivotal positions, hence the sensitivity, to change. Thus the sensitivity is not linear but follows a logarithmic law, being at a maximum for quiet sounds reducing to a minimum for loud ones. This 'automatic volume control' allows the ear to handle an enormous range of sound levels of the order of 10^{12}:1.

Variations in atmospheric pressure on the front of the ear drum would disrupt the delicate action of the ossicles unless compensated by equal pressure on the back. This is accomplished by the *eustachian tube* which connects the middle ear to the back of the throat.

Frequency division   The inner ear consists of a long tube filled with fluid and rolled up like a shell, called the *cochlea*. A horizontal *basilar membrane* divides the tube along its length into upper and lower compartments except at the sealed far end where there is a connecting gap called the *helicotrema*. The compartments are termed the *scala vestibuli* and the *scala tympani* respectively.

Sound vibrations are applied to the oval window at the entrance of the upper chamber by the stirrup bone. From there they travel through the fluid to the gap at the end, down into and along the lower compartment and back to its round window, where they are absorbed. En route they pass through thousands of sensitive hair cells on the upper surface of the membrane, which are linked to nerve fibres.

These cells respond to different frequencies and are divided into 24 bands with one third octave spacings, starting with the highest band near the entrance and the lowest at the far end. Individual bands occupy about 1.3 mm of space, each being termed a *bark*.

Centre frequencies of the bands start at 50 Hz for band 1 and go up to 13.5 kHz for band 24. Cut-off outside each band is sharp at the lower side but more gradual at the high. There is thus an overlap of adjacent bands. The lowest band has a bandwidth of 100 Hz and the highest 3.5 kHz. Overall response in a healthy person under 30 years old is 16 Hz–16 kHz.

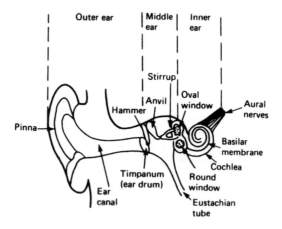

Figure 1. Diagram of human ear showing outer, middle and inner sections.

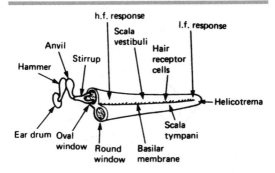

Figure 2. Illustration of the basic components of the inner ear showing the cochlea straightened out to reveal its various features.

Frequency response   Response of the ear is not flat but is at a maximum from 2–4 kHz. The rest of the curve varies according to the sound level. At lower volumes, response to both treble and bass is less which is why some audio units boost these at low listening levels.

Speech frequencies come well within the overall range unless the hearing has been impaired. Music encompasses a greater range both in loudness and frequency.

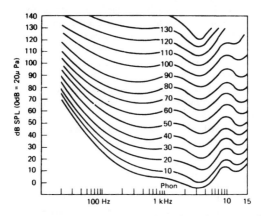

Figure 3. Equal-loudness contours. These show the amount of sound pressure required to produce sensations of equal loudness at various frequencies and volume levels. They are therefore the inverse of frequency response curves. The contours are an average over a large number of persons aged 18–25.

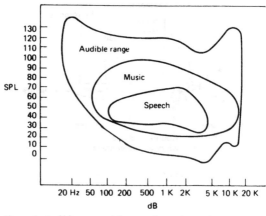

Figure 4. Audible range of frequencies and sound pressure levels. The threshold of hearing is the minimum that can be heard, while the threshold of feeling produces a tickling sensation in the ears. Regions for music and speech are shown.

Hearing damage   Temporary reduction of hearing sensitivity, as indicated by the increase of hearing threshold level, occurs after exposure to loud sounds. This can become permanent if the exposure is prolonged. The louder the sound, the shorter the exposure time to produce permanent impairment. Damage is greater if the sound contains impulsive components caused by percussive elements in the source.

Hearing impairment is of a band centred around 4 kHz (bark 18 in the cochlea), irrespective of the nature of the sound causing the damage. As damage increases with further exposure, the band broadens, reaching in some cases down to 1 kHz.

Exposure levels and times are regulated in industry. Maximum permissible levels are shown in the following chart:

dBA	Time	dBA	Time
90	8 hours	102	30 minutes
93	4 hours	105	15 minutes
96	2 hours	108	7.5 minutes
99	1 hour	111	3.75 minutes

It should be noted that with disco music and headphone listening, levels well in excess of 100 dBA are possible. The danger is therefore obvious.

Presbycusis Hearing ability deteriorates with age, a condition known as presbycusis. It starts slowly from 20–30 years of age, but worsens rapidly in later years. Ageing may not be the sole cause, but exposure to the noises of everyday civilisation could play a part.

This has been dubbed 'sociocusis'. Measurements of the hearing thresholds of nomadic tribesmen living in exceptionally quiet conditions revealed a much lower rate of hearing loss with age, although other non-acoustic factors could be involved.

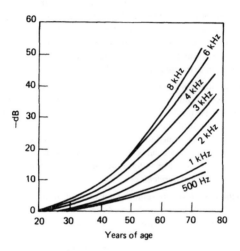

Figure 5. Age-related hearing loss. Different surveys have been made that broadly correlate although differing in minor detail. These curves are of an average of these surveys. Frequencies affected are from 500 Hz upward, those below showing little deterioration.

Sound sources

Sound in air behaves in a similar way to ripples in a pond except it radiates in three dimensions instead of two and the waves are lateral rather than perpendicular to the plane of propagation — they oscillate back and forth instead of up and down. Thus alternate areas of high and low pressure are generated, one cycle consisting of one of each.

Any disturbance in the air causes pressure waves to radiate outward; if these or any component of them lie within the frequency range of human audibility they are classified as sound waves. Those lying below that range are termed *infrasonic*, while those that are higher are *ultrasonic*.

Pure tones consist of a single-frequency and the wave is said to be a *sine*, or *sinusoidal* in character. The air particles move with a simple harmonic motion like a pendulum, although in a straight line instead of in an arc.

Fourier analysis　Few natural sources produce a pure tone; most have several radiating surfaces, each vibrating in different modes. These result in waves of irregular shape, though if the sound is sustained, the successive waves are similar to each other and so are termed *periodic*. By means of Fourier analysis these irregular shapes can be analysed as consisting of a fundamental sine wave with various amounts of multiples which are termed *harmonics*.

Harmonics　Most conventional musical instruments produce sound from a vibrating string or air column. This is acoustically amplified by means of a sounding box in the case of the strings, or a flare which increases the air-coupling efficiency with most of the wind instruments.

A string vibrating at its resonant frequency forms an ellipse of vibration between two null points or *nodes* with a maximum or *antinode* in the centre. Simultaneously, both halves vibrate generating a harmonic at twice the frequency; also thirds vibrate at three times the frequency and so on. Each set of vibrations has its nodes and antinodes.

The same applies to wind instruments, but the location of the nodes and antinodes in the resonating column of air depends on whether the pipe is closed or open at one end. With an open pipe, the node appears in the middle and the antinodes at both ends, but a pipe that is closed at one end has its node at the closed end and one antinode at the open. The harmonics follow a similar pattern to the fundamental, the open pipe having antinodes at both ends, and the closed pipe having one node at the closed and an antinode at the open end.

Flares Most wind instruments have some sort of flare at the free end which gives the most efficient coupling known of a small high intensity source to the free air. The efficiency is related to the length of the horn and width of the flare compared to the wavelength of the sound.

Low frequencies thus require large dimensions. With brass instruments the required length is obtained by coiling the tube in a practical configuration. The low frequency response required of a full-range horn loudspeaker makes the size a major problem.

There are various flare shapes, the simplest and least effective being conical. The most common shape follows a logarithmic or exponential curve. Another type is the hyperbolic.

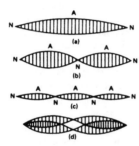

Figure 6. Vibrating string showing nodes (N) and antinodes (A): (a) fundamental; (b) second harmonic; (c) third harmonic; (d) how string vibrates to more than one frequency simultaneously (fundamental and second, harmonic).

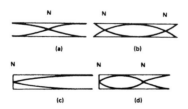

Figure 7. Air column nodes and antinodes: (a) open-ended, fundamental; (b) second, harmonic; (c) closed one end, fundamental; (d) second, harmonic.

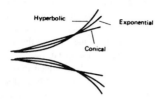

Figure 8. Flare shapes.

Percussive sounds   Produced when one object strikes another, these occur naturally as well as in music. They are characterised by a sudden rise to maximum amplitude termed a *transient*, followed by a slower, though still rapid decay. The nature of the source will in many cases produce frequencies not harmonically related to the fundamental and successive waves have no similarity. They are said to be *aperiodic*. Percussive sounds having a recognisable pitch such as those made by a bell, tympani or a piano, are an exception.

Frequency response   Each octave in the musical scale doubles the frequency of the previous one although the intervals sound the same. The fundamental orchestral range is from 27.5 Hz–4,186 Hz, but the harmonics, which are multiples of each note, extend the upper range to the limit of audibility. These, along with starting transients, give the instruments their characteristic sound.

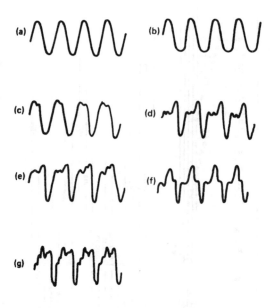

Figure 9. Different harmonic content of various instruments:
(a) glockenspiel; pure tone with very little harmonic content;
(b) piano; soft, mainly second harmonic (broader negative
half-cycles); (c) piano; loud. other harmonics, mostly even,
appear; (d) trumpet; strong harmonics giving bright tone;
(e) french horn; fewer harmonics than trumpet producing a
mellower tone; (f) clarinet; large number of odd harmonics
with strong fundamental; (g) violin; large number of odd and
even harmonics.

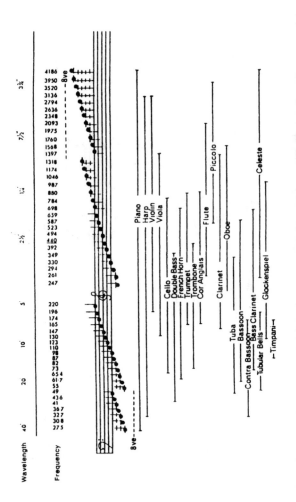

Figure 10. Fundamental frequency range of orchestral instruments. Fundamentals are normally strong with weak harmonics, but in the lower bass region, the fundamentals are usually weak and less significant with the second and third harmonics being dominant. Thus reproducers with a poor response over the lowest octave do not sound greatly inferior.

Measuring sound

The decibel   As aural perception of different sound intensity
levels follows a logarithmic law, the method of measuring
them is also logarithmic. The unit is the decibel (dB) being a
tenth of a Bel; it expresses a ratio between two values and so of
itself is not absolute.

Sound intensity level   Sound intensity is the energy in walls
passing through an area of one square metre perpendicular to
the direction of propagation. Sound intensity levels in dB are
compared to a reference of 10^{-12} watts/m² (0 dB), which is the
faintest sound at 1 kHz that can be heard by a healthy young
adult. This deflects the eardrum by less than the diameter of a
single atom. The formula is:

$$IL = 10\log_{10}\left(\frac{W}{W_r}\right)$$

where IL = level in dB; W = intensity in watts; and
$W_r = 10^{-12}$ watts. Thus a 1 watt intensity = 120 dB.

Sound pressure   The energy transmitted by a wave is
proportional to the square of pressure exerted. As sound
pressure is easiest to measure, it is the parameter most used to
describe sound levels, being thus proportional to the square
root of the intensity. Sound pressure levels are also compared
to the faintest audible sound for which the 0 dB pressure
reference is 20 µPascals. The formula is:

$$SPL = 20\log_{10}\left(\frac{\rho}{\rho_r}\right)$$

Where SPL = sound pressure level in dB; ρ = pressure in µPa;
$\rho_r = 20$ µPa. The word *level* is appended to the terms *intensity*
and *pressure* to denote that they are decibel values referred to
the hearing threshold.

Pressure is now measured in *Pascals*, but other units have
been used. Their relationship is:

10 dyne/cm² = 10 µbar = 1 Newton/m² = 1 Pascal

The *bar* is the atmospheric pressure at sea level, so the Pascal
is 10^{-5} of the atmospheric pressure.

Loudness   The perception of loudness depends on the sound
pressure level and frequency. Sounds at the low or high end of
the spectrum seem less loud than those in the middle region
because of the ear's varying sensitivity at different frequencies.

The phon The unit of loudness is the phon, which takes into account the dependence on frequency of the ear's sensation of loudness. One phon is equivalent to 1 dB *SPL* at 1 kHz. When sound pressure levels are quoted at 1 kHz, they are equal to the number of phons. At other frequencies the number of phons and the dB *SPL* level are different (see Figure 3).

The sone This is the loudness of a 1 kHz tone at a level of 40 dB, hence is equal to 40 phons. It is not often encountered.

Middle frequency In numerous audio measurements and tests the frequency of 1 kHz is used. This is because it is almost the mid-point in the audio frequency spectrum. The actual mid-point is 640 Hz, from which 5 octaves below is 20 Hz, and 5 octaves above is 20 kHz.

Average pressure levels of common sounds

130	
	Jet aircraft taking off
120	
110	Pneumatic drill at 1 m
	Wood working machinery
100	Discos
90	Symphony orchestra playing fff
80	Outdoor p.a. system (peaks)
	Vacuum cleaner at 1 m
70	Inside cruising motor coach
	Speaking voice at 1 m
60	General office
	Orchestral woodwind solos
50	Whisper at 1 m
	Orchestra (strings) playing ppp
40	
	Quiet living room
30	
20	
	Quiet countryside
10	
0	Threshold of hearing

Weighting curves Because the ear is not equally sensitive to all frequencies, weighting curves are applied when measuring sounds in order to reflect their the aural effect. Furthermore, as the frequency response of the ear varies according to the loudness of the sound, different curves have been used for different sound pressure levels.

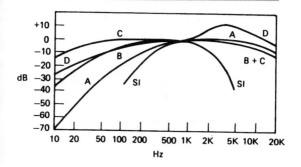

Figure 11. Weighting curves: A for levels below 55 phons; B for 55–85 phons; C for levels above 85 phons; D for measuring aircraft noise accounting for high-frequency whine; SI a suggested curve for assessing speech interference, concentrating on speech frequencies. B and C curves have not compared well with subjective results owing to the use of pure tones for the tests. Hence the A curve is now used for nearly all measurements at any level. The designation for such weighted values is dBA.

Particle velocity The differentiation of the displacement of air particles with respect to time. The oscillatory speed of air particles. Not to be confused with the speed of sound through air.

Specific acoustic impedance The ratio of acoustic pressure to particle velocity. For plane waves it is also the product of the speed of sound in a medium and its density. The unit is the rayl; for air at normal pressure and at $20°$ C the impedance is 415 rayls.

Sound power level The sound power level (SWL) is a measure of the total sound energy output of a sound source. A measurement of the sound pressure at any point does not give a true picture of the output of the source because pressure is affected by the environment; furthermore the source may be directional, radiating more sound in one direction than another. Pressure measurements can thus vary with direction.

Sound power must not be confused with the electrical power used to drive a sound producing device such as a loudspeaker.

The ratio of sound power to electrical power is a measure of the efficiency of the transducer which may be as low as a few per cent.

The formula for calculating sound power level is:

$$SWL = 10\log_{10}\left(\frac{W}{W_r}\right)$$

where SWL is the acoustic power in dB; W is total radiated power, and W_r is the reference power 10^{-12} watts.

Typical sound power levels

dB (10^{-12} W) Power (W)   Power (W)

dB (10^{-12} W)	Power (W)		Power (W)
200	10^8	Saturn rocket	50 000 000 W
180	10^6	Jet airliner	50 000 W
160	10^4		
140	100	Symphony orchestra	10 W
120	1	Chipping hammer	1 W
100	10^{-2}	Shouted speech	0.001 W
80	10^{-4}	Conversational speech	20 x 10^{-6} W
60	10^{-6}		
40	10^{-8}	Whisper	10^{-9} W
20	10^{-10}		
0	10^{-12}		

Decibel ratios   Analogue signal voltages, currents, or powers that represent sound waves, can also be conveniently compared using dB notation, providing the circuit impedance is the same for both values. Thus the aural effect of a specified loss or gain can be determined.

Approximate decibel ratios

Several approximate ratios for sound pressure, voltage or current are worth remembering:

3.5 dB = x 1.5: 6 dB = x 2; 10 dB = x 3: 14 dB = x 5: 20 dB = x 10

If the impedance is not equal in the two circuits under comparison, the gain or loss between two values of current or voltage must be related to the respective impedances.

For current, the formula is:

$$\text{difference in dB} = 20\log_{10}\frac{I_1}{I_2} + 10\log_{10}\frac{Z_1}{Z_2}$$

For voltage, the formula becomes:

$$\text{difference in dB} = 20\log_{10}\frac{V_1}{V_2} - 10\log_{10}\frac{Z_1}{Z_2}$$

Ratios of intensity and sound pressure levels

IL or power ratio	SPL voltage or current ratio	dB	SPL voltage or current ratio	IL or power ratio
1.000	1.000	0	1.000	1.000
0.977	0.989	0.1	1.012	1.023
0.955	0.977	0.2	1.023	1.047
0.933	0.966	0.3	1.035	1.072
0.912	0.955	0.4	1.047	1.096
0.891	0.944	0.5	1.059	1.122
0.871	0.933	0.6	1.072	1.148
0.832	0.912	0.8	1.096	1.202
0.794	0.891	1.0	1.122	1.259
0.708	0.841	1.5	1.189	1.413
0.631	0.794	2.0	1.259	1.585
0.562	0.750	2.5	1.334	1.778
0.501	0.708	3.0	1.413	1.995
0.447	0.668	3.5	1.496	2.239
0.398	0.631	4.0	1.585	2.512
0.355	0.596	4.5	1.679	2.818
0.316	0.562	5.0	1.778	3.162
0.251	0.501	6.0	1.995	3.981
0.200	0.447	7.0	2.239	5.012
0.159	0.398	8.0	2.512	6.310
0.126	0.355	9.0	2.818	7.943
0.1	0.316	10	3.162	10.00
0.0794	0.282	11	3.55	12.6
0.0631	0.251	12	3.98	15.9
0.0501	0.224	13	4.47	20.0
0.0398	0.200	14	5.01	25.1
0.0316	0.178	15	5.62	31.6
0.0251	0.159	16	6.31	39.8
0.0159	0.126	18	7.94	63.1
0.01	0.1	20	10	100
0.001	0.0316	30	31.6	1000
0.0001	0.001	40	100	10^4
10^{-5}	3.16×10^{-3}	50	316	10^5
10^{-6}	10^{-3}	60	1000	10^6
10^{-7}	3.16×10^{-4}	70	3162	10^7
10^{-8}	10^{-4}	80	10^4	10^8
10^{-9}	3.16×10^{-5}	90		10^9
10^{-10}	10^{-5}	100	10^5	10^{10}
10^{-11}	3.16×10^{-5}	110		10^{11}
10^{-12}	10^{-6}	120	10^6	10^{12}

Nepers To convert decibels to nepers, multiply by 0.1151. To convert nepers to decibels, multiply by 8.6881. (Nepers are an alternative unit to the decibel involving some rather obscure maths. It is little used so does not warrant full treatment. A straight conversion from decibels/nepers should suffice for those who may encounter the term and wish to relate it to decibels.)

Sound propagation

Monopole/point source   When a source radiates equally in all directions, the sound pressure areas emanate from it like expanding spheres. The energy is distributed over the surface of each sphere, and so is spread over an increasing area as each expands outward. It thus diminishes as the distance from the source increases, being inversely proportional to the area of the sphere. Sound intensity at any point from the source is thereby given by the formula:

$$I = \frac{W}{4\pi r^2}$$

in which I = intensity in watts; W = source power; r = distance in metres.

Intensity attenuation thereby follows a *square law*, decreasing to a quarter for a doubling of distance, thus giving: $10\log_{10} 0.25 = 6$ dB. Pressure, being the square root of intensity, decreases in *direct proportion*, so halving at double the distance. Using the formula for sound pressure we get $20\log_{10} 0.5 = -6$ dB. Thus the dB values for intensity or pressure are the same.

Sound pressure is derived from a monopole source power *SWL* as follows:

$$SPL = SWL - 20\log_{10} r - 10\log_{10} (4\pi)$$

hence:

$$SPL = SWL - 20\log_{10} r - 11 \text{ dB}$$

where *SWL* is sound power level of source, dB referenced to 10^{-12} watts.

A loudspeaker in a sealed enclosure behaves as a monopole for all wavelengths greater than its own dimensions.

Dipole/doublet   The dipole can be thought of as two monopoles back to back and 180° out of phase. However, the source is at the edge rather than in the centre of each radiation sphere. So the pressure level at any given point depends not only on the distance from the source r, but also on the angle from the main axis. The formula is:

$$p_2 = p_1 \cos\theta$$

where p_1 = pressure measured on axis; p_2 = pressure off axis at angle θ.

Attenuation is 6 dB per doubling of distance in the far acoustic field, but is 12 dB per doubling for the near field, that is within a short distance from the source.

Examples of dipoles are: loudspeaker on open baffle or in vented-back cabinet, tuning fork, cymbals, most percussion instruments except tympani.

Line source   A line source radiating sound through 360° does so in the form of concentric expanding cylinders. The sound intensity per unit length of the source is:

$$I = \frac{W}{2\pi r}$$

where I = intensity in watts; W = source power; r = distance in metres.

Sound intensity is thus inversely proportional to the distance, so doubling the distance gives half or -3 dB intensity and sound pressure levels. This is also true when the line source radiates only over a segment of a circle. The power is concentrated in this segment, but the attenuation with distance is the same. Examples are: column speakers, moving trains, motorways, noisy pipes.

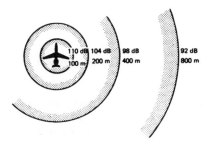

Figure 12. A monopole or point source. Pressure waves form expanding spheres having a pressure decrease of 6 dB per doubling of distance.

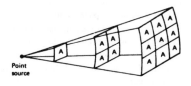

Figure 13. The area covered by a portion of expanding sound wave is the square of the distance from the source. Hence the intensity follows an inverse square law.

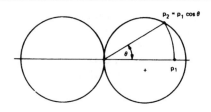

Figure 14. The dipole or doublet. Sound pressure at an angle from the axis is cos θ that of the pressure on axis at the same distance. Attenuation is 12 dB near field, 6 dB far field, for each doubling of distance.

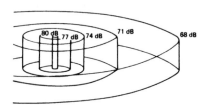

Figure 15. Line source. Attenuation is 3 dB with doubling of distance.

Plane source If a sound wave is confined in a tube with hard walls it is prevented from diverging and so emerges at the end at a pressure equal to that at the beginning minus small losses due to air turbulence and thermal agitation. Most of the energy can thus be transferred over quite lengthy distances. Examples are: acoustic speaking tubes, noisy air-conditioning ducting.

Directivity factor The radiated power of a monopole source depends on its immediate environment. If it is suspended in free air, the power is as shown above. If it is mounted on a wall, floor or ceiling, half of the sphere is blocked so twice the intensity (3 dB) is radiated in the remaining half-sphere of air. If it is placed at the junction of a floor or ceiling and wall, four times the intensity (6 dB) is concentrated in the quarter sphere of air available, while if it is in a corner of floor or ceiling and two walls, the power intensity will be eight times (9 dB) in the remaining eighth sphere.

The increase in intensity compared to a point in a free field is the *directivity factor*, and the decibel equivalent, the *directivity index*.

Source location	Directivity factor	Directivity index
Free suspension	1	0 dB
On flat plane	2	3 dB
Junction 2 planes	4	6 dB
Junction 3 planes	8	9 dB

Speed of sound Sound velocity depends on atmospheric pressure and temperature, the latter being the most significant factor. The formula is:

$$V = 332\sqrt{1 + \frac{t}{273}}$$

where V = velocity in m/s, and t = temperature in °C. Alternatively:

$$V = 1089\sqrt{1 + \frac{t}{273}}$$

where V = velocity in ft/s, and t = temperature in °C.

The velocity at 0° C is 1090 ft (332 m) per second, rising by 2 ft (0.6 m) for each degree C increase in temperature. Thus at 15.5° C (60°F) the speed is 1119 ft (341 m) per second. This is approximately 1 mile in 5 seconds. The distance of a lightning strike can thus be calculated by counting the interval between the flash and the thunder.

Wavelength/frequency The formula for determining the wavelength of a sound wave is:

$$\lambda = \frac{V}{f}$$

in which λ is the wavelength; V = sound velocity; f = frequency in Hz. λ and V must be in the same units, inches, feet or metres. In view of the effect of temperature on V, wavelength will vary slightly with temperature changes.

Near/far fields The near field is that area immediately surrounding the sound source. Sound pressures may vary at different points within the field and not always be directly proportional to the sound intensity. The particle velocity of the air molecules may not be in the direction of propagation.

A precise determination of the extent of the near field is not normally possible as many factors such as frequency, size and shape of the course and parts of the source vibrating at different phase angles from the others, can affect the near field.

In the far field, propagation is more stable although there may be local disturbances due to objects in the field. Here. the particle velocity is mainly in the direction of propagation, and intensity is proportional to the sound pressure. The attenuation is at 6 dB per doubling of distance or 3 dB for a line source. If pressure measurements are taken at various distances from the source. it is where they begin to show the 6 dB pattern of attenuation that the far field can be deemed to commence.

Reverberant field   If the source is radiating in an enclosed area such as a room or hall, reflections from the boundary walls create a reverberant field which is superimposed on the far field.

Sound intensity in a reverberant field is given by:

$$I = \frac{4W(1-\alpha)}{S\alpha}$$

in which W = source output power; S = total room surface area: α = average room absorption coefficient.

Combination of reverberant and direct fields is given by:

$$I = \frac{WD}{4\pi r^2} + \frac{4W(1-\alpha)}{S\alpha}$$

in which D = directivity factor.

The resulting sound pressure level is the sum of the *SWL* (sound power of the source in dBs) and 10 times the logarithm of the above expression without the source power factor:

$$SPL = SWL + 10\log_{10}\left(\frac{D}{4\pi r^2} + \frac{4(1-\alpha)}{S\alpha}\right)$$

Room radius   The closer to the source, the greater the proportion of direct sound to reverberant sound. At a certain distance both will be equal, this point is known as the room radius. It is given by:

$$r = \sqrt{\frac{S\alpha D}{(1-\alpha)16\pi}}$$

Reflection   When meeting a surface, part of the sound wave is absorbed, part transmitted through it and part reflected. If absorption and transmission are low most of the energy is reflected and the surface is said to be acoustically hard. Absorption, hence reflection, varies over the frequency range with most surfaces.

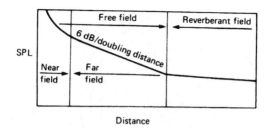

Figure 16. SPL variation in room or hall indicating the three acoustic fields.

The angle at which a sound is reflected is the same as the angle of incidence, when compared to a perpendicular line raised from the surface. When a corner is encountered, the wave is reflected back on a separate but parallel path. A concave surface focuses the sound, while a convex one disperses it. Thus sound is reflected similarly to light.

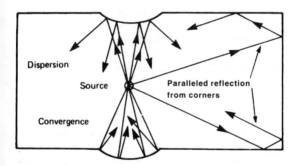

Figure 17. How differently shaped surfaces reflect sound. Sound reflected from a flat surface appears to have a ghost image behind the surface as its source. Where the reflected high pressure regions coincide with those of the original wave, reinforcement occurs, but where they cross the low pressure areas, cancellation takes place. Sound pressure measurements can thus be inaccurate in the presence of hard reflecting surfaces.

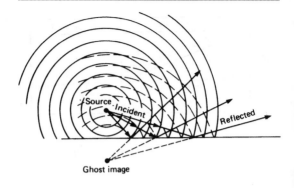

Figure 18. Reflections from a flat surface showing ghost image.

Diffraction According to Huygen's theory of wavefronts, each point of a wavefront is considered a new point-source radiating in all directions. However it maintains a forward direction because of the pressure of adjacent particles inhibiting side-spread. But when a wave encounters an obstruction, there is no side pressure beyond its edge, so the waves spread behind it as if the edge was a new source.

Similarly, when a wavefront encounters a hole in a surface, it spreads through the hole as if the hole was a source. A plane wave (an almost flat wavefront from a distant source) thus becomes spherical.

The energy of the new waves is that of the originals appearing over the area of the hole. Sound pressure is thus considerably less than the original and dependent on hole size.

The above is true when the wavelength is large compared to the obstruction or hole size. When it is relatively small (at high frequencies), a shadow is cast behind an obstruction, and sound is beamed through the hole.

Refraction Sound waves are bent when they pass from a medium of one density into one of another, owing to their differing sound velocities. In air, refraction occurs with either a *temperature* or *wind gradient*.

Temperature normally decreases with height, so sound has a greater velocity at the warmer ground-level region than at a higher point. This turns sound upward and the turn continues as long as decreasing temperatures are encountered, i.e. throughout the temperature gradient.

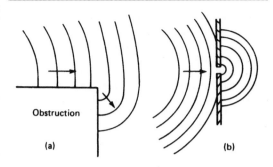

Figure 19. (a) long wavelengths are diffracted behind an obstruction, the edge behaving as a new source; (b) hemispherical diffraction of long wavelengths through a hole in a surface.

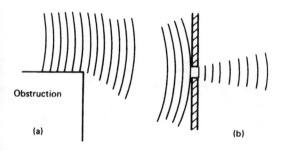

Figure 20. (a) shadow caused by short wavelengths behind an obstruction; (b) short wavelengths beaming through a hole.

Sometimes there is a *temperature inversion* whereby the gradient is positive and the temperature increases with height up to a certain level. This bends the sound waves downward to reinforce those propagated parallel to the ground and so they carry further. The effect often occurs over a stretch of water.

Rarely, there is a *double gradient*, with the temperature decreasing with height to a certain point, then increasing to a further point beyond which the normal decrease again takes over. Sound waves curve upward through the first gradient, are bent back by the second, are turned upward again by the first, and so on. Sounds can thereby carry over a considerable distance with little attenuation.

Wind velocity increases with height above the ground thus establishing a gradient. Sound propagated against the wind curves upward, while that travelling with the wind is refracted downward. This is not due to the wind itself but the wind velocity gradient.

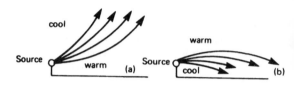

Figure 21. (a) refraction with a normal temperature gradient; (b) refraction with an inverted temperature gradient.

Figure 22. Refraction with double gradient. Sound is trapped between the gradients and can travel considerable distances.

Effect of wind   Wind velocity is zero at actual ground level and increases with height. Sound wave radiation direction is bent toward the higher velocity layers which is upward, when propagated against the wind; and away from those layers or downward when propagated with the wind. Sound is thereby attenuated in the direction of the wind and reinforced in the opposite direction.

Effect of humidity   The presence of moisture in the air affects sound propagation. In dry air attenuation is low, but it increases to a critical region between 15–20% relative humidity. Above this, it decreases to a minimum at high

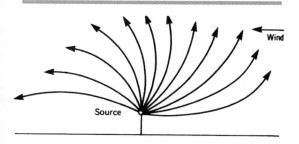

Figure 23. Sound refraction caused by wind.

humidity levels. Propagation is thus at its greatest in mist or fog, aided also by the still air usually present in those conditions.

The effect is frequency-dependent, with attenuation being high at high frequencies. The frequency differential is especially marked at the critical relative humidity.

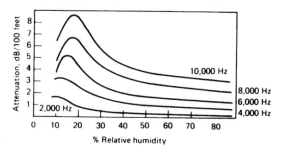

Figure 24. Attenuation of sound at various relative humidities.

Frequency and energy transmission   When a sound wave passes through a 'slice' of air of particular thickness, there is a difference in pressure between the front of the slice and the back due to the changing pressure levels along the wave. This difference accelerates the air molecules and energy is transmitted.

At half the frequency a wave takes twice the time to change pressure, so it is only half as steep. Thus the pressure difference across the same slice of air is also just a half of the original. However, the pressure is sustained for twice as long, so the total force accelerating the air particles is the same, hence is frequency independent.

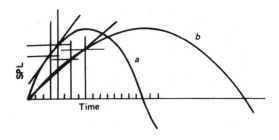

Figure 25. (a) pressure gradient over a slice of air, (b) gradient is halved at half the frequency, but duration is twice as long. Total energy is the same.

Atmospheric absorption   In addition to attenuation due to the divergence of sound waves, either according to inverse square or inverse proportion depending on the nature of the source, there is a small loss resulting from the absorption by the air itself.

This is insignificant at short to medium distances, but becomes important over long ones. It is frequency-dependent, with the higher frequencies sustaining greater attenuation.

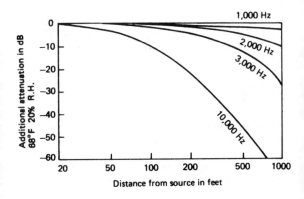

Figure 26. Attenuation due to atmospheric absorption.

Room and hall acoustics

The acoustic of a room or hall describes the characteristic sound resulting from the multiple reflections, absorptions, and resonances at many different frequencies, due to the shape, size, location and material of the various surfaces in the hall, as well as the shape and size of the hall itself. It can vary from one part of a hall to another and is highly complex. No two halls are identical, and determining the acoustics by measurement is a long and laborious task.

Reverberation time   This is the time taken for sound pressure from multiple reflections to die away to a thousandth or –60 dB of its original value and is a major factor in determining acoustic properties. The decay is exponential, but when measured in decibels is linear. As it is rarely possible to measure a decay of 60 dB, a smaller drop of 40 or even 20 dB can be taken and the time multiplied accordingly.

Reverberation is responsible for giving body, fullness and volume. Without it, music and speech sound weak and thin, as in the open air. However too much reverberation causes speech syllables to run into each other and makes music muddled. There is an optimum level which depends on the use and the

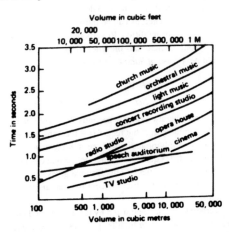

Figure 27. Optimum reverberation times for various uses and sizes of auditoria, studios and domestic listening rooms. Times shown are overall, as they vary for different frequencies. The time at 100 Hz is generally about twice that at 5 kHz, but halls differ in their frequency/reverberation pattern.

size of the hall. Speech requires less than music, and music-recording studies need less than concert halls because the reverberation of the listening room is added to that on the recording.

Test for speech clarity   As normal words could be guessed if misheard, a speaker reads monosyllabic nonsense words from the platform, with members of an audience writing down what they think they heard. The accuracy gives *a percentage articulation index* (PSA). Maximum is usually 95%; 80% indicates good audience comprehension; 75% means the audience needs to concentrate, while below 65% means poor intelligibility.

Absorption   Reverberation time and frequency pattern is governed by the amount of absorption present. All materials absorb some sound and reflect some; rooms having highly reflective surfaces are said to be *reverberant*, a characteristic which is increased by making all the walls non-parallel. These are used for special tests. At the opposite end of the scale are rooms designed to reflect very little sound. These too are used for testing equipment and are termed *anechoic*.

Absorption is a function of the area of a material and its absorption coefficient. To achieve a desired reverberation time, suitable absorbent materials can be introduced, or existing material removed or modified. Reverberation times can be calculated from the materials present and their area, and the room volume, by use *of Sabine's formula:*

$$T_r = \frac{0.161V_m}{S\alpha} \quad or \quad T_r = \frac{0.05V_f}{S\alpha}$$

where T_r = reverberation time in seconds; V_m = volume in cubic metres; V_f = volume in cubic feet; $S\alpha$ = total absorption obtained by multiplying individual areas of the materials (square metres first equation; square feet second equation) by their absorption coefficients. This formula gives results consistent with measured times except for near anechoic conditions.

Reverberation can be adjusted by the amount and choice of materials to favour low or high frequencies, giving a warmer or brighter effect respectively. A high coefficient means high absorption hence low reverberation at that frequency.

Absorbers   Auditoria and studios need different acoustic treatment than domestic listening rooms where furnishings usually provide sufficient absorption. Apart from adjusting reverberation time, resonances at specific frequencies need to be controlled.

Typical absorption coefficients

Material	125 Hz	250 Hz	500 Hz	1 kHz	2 kHz	4 kHz
Acoustic panelling	0.15	0.3	0.75	0.85	0.75	0.4
Acoustic tiles $^3/_4$ in thick $^3/_{16}$ holes, $^1/_2$ centres	0.1	0.35	0.7	0.75	0.65	0.5
Brick	0.024	0.025	0.03	0.04	0.05	0.07
Carpet, thin	0.05	0.1	0.2	0.25	0.3	0.35
Carpet, thick with underlay	0.15		0.35		0.5	
Chair, padded dining*	1.0		2.5		3.0	
Chair, upholstered*	2.5	3.0	3.0	3.0	3.0	4.0
Concrete	0.01	0.01	0.02	0.02	0.02	0.03
Curtains	0.05	0.12	0.15	0.27	0.37	0.5
Curtains, heavy with folds	0.2		0.5	0.6		
Door, wooden*	0.3		0.15		0.05	
Fibreglass, 1 in	0.07	0.23	0.42	0.77	0.73	07
Fibreglass, 2 in	0.19	0.51	0.79	0.92	0.82	0.78
Fibreglass, 4 in	0.38	0.89	0.96	0.98	0.81	0.87
Floor, wood on joists	0.15	0.2	0.1	0.1	0.1	
Floor, wood blocks	0.05		0.05		0.1	
Glass	0.03	0.03	0.03	0.03	0.02	0.02
Person, seated*	0.18	0.4	0.46	0.46	0.5	0.46
Plasterboard	0.3		0.1		0.04	
Plaster on lathe	0.3		0.1		0.04	
Plaster on brick	0.024	0.027	0.03	0.037	0.039	0.034
Plywood $^3/_8$ inch	0.11		0.12		0.1	
Plywood, $^3/_{16}$ on 2 in batten	0.35	0.25	0.2	0.15	0.05	0.05
Wood, $^3/_4$ solid	0.1	0.11	0.1	0.06	0.06	0.11

All are per m² or per ft², depending on the equation used, except those marked * which are for one item.

1.0 = 100% absorption (an open window); 0 = 100% reflection.

Absorption materials will strongly absorb a particular frequency when supported $^1/_4$ wavelength from a wall. The frequency corresponding to $^3/_4 \lambda$, will also be strongly absorbed. Percentage absorption can be as high as 85% compared with 10% at all other frequencies.

Thickness of material determines low frequency absorption.

1 inch	2 inch	3 inch	4 inch	5 inch	6 inch
25 mm	51 mm	76 mm	102 mm	127 mm	152 mm
2.2 kHz	1.4 kHz	800 Hz	550 Hz	450 Hz	400 Hz

Different thicknesses of felt on a solid surface absorb uniformly down to the frequencies shown, below which absorption falls off at an approximate rate of 30% per octave.

Panel absorbers   If a panel is supported by its edges away from a wall, it will vibrate at its resonant frequency thus absorbing energy at that frequency. However, it re-radiates thus reducing its effectiveness as an absorber to a coefficient of less than 0.5. Introducing damping material between the panel and wall increases absorption and broadens the band of frequencies affected. The resonant frequency of a panel is found by:

$$f_r = \frac{6000}{\sqrt{md}}$$

where m = mass of the panel in kg/m^2; d = distance from plate to wall in metres.

Helmholtz resonators   These are resonant chambers which are open to the air at one end through a small neck. Some have their volume adjustable by a plunger like a car foot pump. They absorb energy from a sound field at their resonant frequency, but some is re-radiated after the incident sound has ceased. Thus the amplitude is decreased but reverberation time may be prolonged.

   Because of their sharp frequency response they are used only to tame single frequencies. Response can be broadened by lining the chamber with absorbing material, but this also reduces their effectiveness. The resonant frequency can be calculated from:

$$f_r = \frac{340}{2\pi} \sqrt{\frac{S}{lV}}$$

where S = cross-section area of neck in m^2; l = length of neck in m; V = volume of chamber in m^3.

Perforated panels   If a panel having a pattern of holes in it is mounted with a gap between it and a wall, each hole and the gap behind it will behave as a small resonator. If the holes are of the same size and spacing, they will all resonate at the same frequency but the band is broader than that of a single helmholtz resonator. Further band broadening is achieved by filling the gap with mineral wool or fibre glass. Varying the hole size and spacing gives yet more broadening of the band. Resonant frequency depends principally on the depth of the gap: the deeper the gap the lower the frequency.

Membrane absorbers   If a membrane such as roofing felt is placed between the perforated panel and the filling, the resonant frequency is lowered. A double layer reduces it further. This can be more convenient and space saving for low frequencies than increasing the gap depth. Panels can be made free standing with a solid back-plate and thus be placed at critical points in a studio.

Assisted reverberation   Requirements conflict when halls are used for music and speech. One solution is to design for short reverberation for speech, then position helmholtz resonators containing microphones to receive reflected sound. The delayed sound in the resonators is amplified and fed to speakers in the auditorium. Thus reverberation can be lengthened as required for music.

Room resonances   In addition to reverberation, air resonances greatly affect the acoustics of a room, There are three main ones, the half-wavelength of each corresponding to the dimensions of length, width and height. These fundamentals are accompanied by second, third, etc. harmonics. The frequencies can thus be calculated by:

$$f = \frac{V}{2 \times d}$$

where V = velocity of sound; d = room dimension in same units as V.

Between two opposite room boundaries, sound pressures are constant at specific positions at the resonant frequency. They are therefore known as *standing waves*. For the fundamental, pressures are low at the centre and high at the boundaries, these points being termed the node and anti-nodes respectively. Harmonics have a series of nodes and antinodes, but antinodes always occur at the boundaries. For large halls the fundamental resonances are too low to be excited, but harmonics can be troublesome.

Room corners thus have high SPL values for every mode of room resonance; furthermore a sound source such as a loudspeaker placed in a corner will excite every mode. The room centre has the fewest antinodes and so is the best point acoustically (if not practically) for locating the source, to avoid exciting resonances.

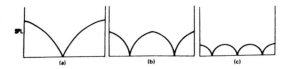

Figure 28. SPL values between opposite walls at resonant frequency: (a) fundamental; (b) second harmonic; (c) third harmonic.

Golden ratio   If two dimensions are similar (a square room), the resonances will reinforce each other and give a strong emphasis to all sounds of that frequency. A room that is a cube is even worse.

Ideally, room dimensions should be such that all resonances and harmonics of each are well spaced. The golden ratio to achieve this is 1:1.25:1.6. Any dimension could be a multiple of one or more of these, i.e. 1:2.5:3.2.

Standing waves can be reduced by fitting absorbents on one of the opposite pairs of walls, but not so much as to excessively reduce reverberation time. Non-parallel walls and bay windows reduce resonance effect so are desirable listening-room features.

Flutter   If the source and the listener are located between two opposite walls which are reflective, while nearby surfaces are absorbent, sound energy will bounce back and forth and decay slowly. The effect is that the sound seems to flutter.

Echo   If a direct sound and strong reflection, or identical sounds from two sources such as loudspeakers, arrive more than 0–05 seconds apart, they are heard as separate sounds, or as an echo. This is the case if the difference between the distance from the listener to each source is more than 50 ft (15 m). Sounds with an interval of less than 0.05 seconds appear as a single, stronger sound coming from the nearer source. This is known as the *Haas* effect.

Reflectors   In large auditoria, sound may be weak and too reverberant because the audience receives insufficient direct sound. Panels of reflecting material can be suspended over and at the side of the stage and in the ceiling of the auditorium at angles to give strong single reflections into all sections of the audience.

To be most effective these should be at least 30 times larger than the longest wavelength. At 10 times larger, some reflection plus some diffraction occurs. When less than 5 times larger, all the sound at that wavelength is diffracted.

Such auditoria, though good for live unamplified performances are poor for amplified public address, because sound from the auditorium is reflected back to the microphones on stage to cause feedback.

Sound insulation

Sound insulation is required to reduce unwanted sounds entering a building from outside, to prevent sounds from one area of a building (noisy workshop, adjacent flats) passing to others, or to create a low noise area (recording studio) within a building.

Sound travels via many paths, blocking some has little effect, all must be identified and dealt with. *Air gaps around windows, doors, through ventilation ducts are a major factor and should be attended to first.*

Materials   On encountering a denser material from air, some sound is reflected, some absorbed, and some transmitted through it. Materials that are good absorbers are generally not good insulators of airborne sound, while good reflectors usually are.

Source propagation   Much depends on how the sound is propagated from the source. It can be mainly airborne or structure-borne. Airborne sounds, such as traffic, radios and conversation, encounter walls which reflect back and absorb much of the sound energy. They thereby serve as sound barriers. When the source is in physical contact with the building structure such as heavy machinery, hi-fi speakers on a bare floor or footsteps, the walls and floor actually propagate the sound. Attenuation of sound through structural materials is small, examples are:

Material	steel	brick	concrete	wood
Attenuation (dB per 30 m)	0.3–1.0	0.4–4.0	1.0–6.0	1.5–10.0

Thus structural borne sound will readily propagate throughout a building, so the source should be isolated from the structure. Machinery can be mounted on a floating platform supported by fibreglass, or heavy carpet used in domestic situations.

Wall insulation   The insulation afforded by a wall to airborne sound is termed its *sound reduction index;* it is expressed in dBs and is the amount by which the sound level on the receiving side is reduced. Sound also enters the receiving room via flanking walls thus decreasing the reduction index.

To take this into account the *apparent sound reduction index* is used. It is defined as:

$$R = 10\log_{10}\left(\frac{W_1}{W_2}\right)$$

where W_1 is the sound intensity incident to the wall; W_2 is the intensity transmitted into the receiving room.

To determine the sound reduction index from sound pressure measurements, the area of the wall and the absorption in the receiving room must be taken into account. The formula is:

$$R = p_1 - p_2 + 10\log_{10}\left(\frac{S}{A}\right)$$

where p_1 = the average sound pressure measured on the incident side of the wall; p_2 = average pressure on receiving side; S = area of wall; A = areas of reflecting surfaces in receiving room multiplied by their absorption coefficients.

Sound reduction characteristics　The reduction of sound by a solid wall is frequency-selective. At low frequencies the stiffness of the material is the main factor; just above these, various resonances cause major variations in sound transmission. At about an octave higher than the lowest resonance, the mass of the wall comes into play. This is termed the *mass controlled region* in which the sound reduction index rises at about 6 dB per octave up to the *critical frequency*.

Three types of wave are commonly present in a solid wall, *compressional*, (as in air); *transverse*, (up and down or side to side motion); and *bending waves*, (ripples). Unlike compressional waves, the velocity of bending waves increases with frequency. So the wavelength of the bending wave is different from that of the incident sound wave which produced it at all except one frequency. This is the critical frequency at which they coincide. Here reinforcement occurs which increases sound transmission thereby lowering the sound reduction index.

As the wavelength of incident sound striking the wall varies with its angle of incidence, the critical frequency also so varies. It is lowest when the angle is greatest, i.e. at grazing incidence of $90°$.

Above the critical frequency, stiffness again takes over to give a further though less steep increase in reduction index. The coincidence effect usually occurs between 1–4 kHz which is the ear's most sensitive region.

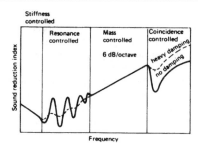

Figure 29. Sound reduction index curve showing the various modes of transmission and the effect of damping.

Damping Damping can be achieved by a thick layer of mastic-like material on one side of the wall, but its effect is confined to the resonance and coincidence modes; it has no effect on the mass-controlled region of the curve. Even so its effect is minimal when a wall has a large mass. Thus it is effective with partitions, but much less so with stone or brick walls.

Mass law Sound reduction in the mass-controlled section can be calculated by the mass law.

$$R = 20\log_{10}(fM) - 47 \text{ dB}$$

where f = frequency of the incident sound; M = surface density of the wall in kg/m^2, (which is the mass in kg/m^2 for homogeneous materials).

Physical characteristics For optimum sound reduction a wall should have low stiffness thus maximising the range between resonance and coincidence, and high mass. Sound reduction increases 6 dB for each doubling of mass, which for a given material is a doubling of thickness.

Cavity walls As losses occur at each interface, a cavity wall having four surfaces has greater sound reduction than a single one having two. The springiness of the air between the cavities reacts with the mass of the leaves to produce resonance. The resonant frequency should be below 100 Hz which is achieved by either a wide cavity (1 m or more) or leaves of large mass, or partly by both. In addition there is a cavity resonance at about 250 Hz which reduces sound insulation.

These resonances can be considerably reduced by introducing absorbent lining such as fibreglass into the cavity. It is especially useful where practical considerations limit the thickness of the wall thus raising the lower resonance above 100 Hz.

The sound reduction of a cavity wall is limited by the necessary edge, tie and footing connections, as these form leakage paths to conduct sound between the leaves. Any practical construction which can minimise or even eliminate these will greatly improve the sound reduction index.

Flanking walls   As these conduct sound into the receiving area, there is little point in improving the insulation of the intervening wall beyond a certain level unless the flanking walls can also be isolated structurally.

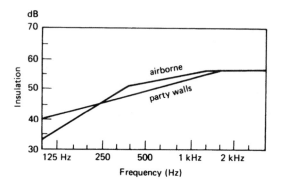

Figure 30. Recommended sound reduction curve for party walls between houses in the U.K.

Ceilings   False and suspended ceilings can reduce overhead noise but are not very effective as the mass required to make an efficient barrier would make them impractical. Heavy carpeting in the area above is the best solution as this also reduces the transmission of sound to the flanking walls.

Sound can pass through the air above suspended ceilings from one area to another, so bypassing dividing walls. The only remedy is to extend the walls above ceiling level to block the air passage.

Floors   Floor insulation can be achieved by installing a floating floor. The structural base is covered with glass fibre, followed by a layer of roofing felt, on which a concrete slab is laid. There must be no physical contact by fixing or skirting board between the building structure and the floor. Where the base consists of wooden joists. a layer of fibre glass is laid over them, then wooden battens are placed on top to match the joists and the floor boards screwed to these.

The floor will have a resonant frequency at which sound reduction is zero. Above this, sound reduction rises at a rate of some 12 dB per octave for low damping to around 6 dB for higher damping. Below the resonant frequency the reduction becomes negative, that is sound transmission is *greater* than if the floor was not floating. Here, the higher damping reduces transmission the most.

So to get a high reduction over the audible range, it is best to have a floor with a very low resonant frequency below 20 Hz and use a small amount of damping. A 3 cm layer of fibreglass with a thick concrete slab is recommended.

Doors Air gaps, including letter boxes and keyholes, are a major source of sound leakage, and should be eliminated. Mass is small compared to a wall, so the resonant frequency is high and there is virtually no mass-controlled region between resonance and coincidence. Doors, however, can be damped by heavy bituminous cladding on one side. Where a high degree of noise reduction is required, two doors both with air seals and separated by a small vestibule is the most effective measure.

Windows Most outside noise comes through window glass. A single pane has a sound reduction index of about 25 dB. With double glazing, a large gap is needed to avoid a high

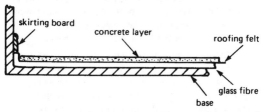

Figure 31. A floating concrete floor laid on a solid structural base.

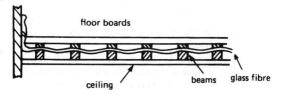

Figure 32. Wooden floating floor on joists.

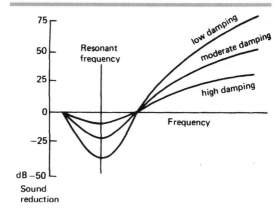

Figure 33. Sound reduction curves of a floating floor. Low damping gives best results above resonance but worst below it. It should therefore be used with a low resonant frequency of below 20 Hz.

resonance frequency owing to the low mass of the glass. Exact figures depend on weight and size but a 4 inch gap is about the optimum, though this is too large for good thermal insulation. Normal gaps of around $\frac{1}{2}$ inch produce resonances in the region of 300 Hz which reduce the sound insulation at that point. Damping the resonance by putting absorbent material within the cavity at the edges of the panes can improve matters. Standard double glazing gives about 4 dB improvement over single panes. It is best to have the panes of different thicknesses, i.e. 6 mm and 4 mm or 10 mm and 6 mm

Air-conditioning/ventilation systems Because plane waves are involved, noise can be carried far from the source. For filtering low frequency noise from the fan or conditioning equipment, a *plenum chamber* which is a large chamber containing baffles and lined with absorbent material, is interposed between the equipment and the ducting. Flexible ducting is fitted to its inlet and outlet to isolate mechanical vibration.

Smooth bends and gradual changes of cross section are used in the ducting system. Internal vanes are fitted to reduce air turbulence, and the inside is lined with absorbent material having a smooth surface.

Noise can be caused by turbulence at the exit grilles, so these are of a type that produce minimum air flow disturbance. Most of those measures should be taken at the design stage, but some idea of the problems and their solution can be useful if noise problems arise later.

Microphones

Microphones can be classified in two ways, the type of transducer and the type of acoustic characteristic. Each has its practical application, some being more suited for one purpose than another.

Transducers

There are several ways of converting sound into electrical energy. The main factors to be considered when choosing a particular type are: fidelity, sensitivity, robustness, convenience in use and cost. Appearance may also be important in some cases, but most types are available in a variety of housings so this should not affect the choice of type.

Moving-coil   A shallow plastic cone has a coil wound on a former at its apex. One magnet pole-piece surrounds the coil while the other is located inside it. The magnetic flux thus cuts through the coil windings, inducing an electromotive force in them when the cone is moved by sound pressure waves. The construction and action is similar to that of a loudspeaker working in reverse.

Fidelity is good, but the mass of the coil and cone produce a resonance between 2 5 kHz giving a peak in the response of several dB. Some makers euphemistically describe this as a *presence effect* because it emphasises ambient noises which tend to be in this frequency band. It also gives a 'bright' tone which is liked by many users and actually does seem to enhance the sound of some musical instruments such as the piano.

The peak can crispen speech of poor intelligibility due to deficiency in dental or fricative consonants, but will make a lisp or similar speech defect sound worse. When used in public address systems, the peak will initiate early acoustic feedback and so reduce the amount of usable gain.

In more expensive models the peak is well damped, but is not completely eliminated; in cheaper ones the peak is usually severe. In general the frequency response is good, but the mass and consequent inertia of the cone restricts the upper response compared to some other types.

Some models have a double cone, one handling the low frequencies and the other the highs, with a crossover circuit to combine the outputs. With these, the small size and mass of the treble cone pushes its peak high out of the mid-frequency region, while that of the bass cone is above its crossover frequency. Thus a response is achieved which is substantially flat over most of its range.

Sensitivity of moving-coil instruments is high. The basic transducer is of low impedance, being around 30 Ω, so the output voltage is also low, however, built-in transformers give outputs of 200 Ω, 600 Ω or 50 kΩ with correspondingly higher voltages.

Robustness is the greatest for any microphone and moving-coil units will stand up to a lot of hard use and even abuse. This also makes them very convenient to use as few precautions are necessary to protect them. They are thus particularly suitable for outdoor or travelling use.

Cost covers a wide range from quite cheap models to very expensive. Flatness of frequency response, hum-shielding, quality of material and construction all affect the cost.

Ribbon   A narrow aluminium ribbon, corrugated to make it flexible, is suspended edgewise between the poles of a magnet. As there are small gaps between the edges of the ribbon and the pole pieces, the ribbon is impelled by the velocity of the air particles as they flow past it rather than sound pressure acting upon it. It is therefore also known as a *velocity* microphone. As the moving ribbon cuts the magnetic lines of force, an electromotive force is induced.

Fidelity is of a high order because the mass of the ribbon is low. Thus it has a low inertia and so responds well to high frequencies. Also the ribbon resonance is high, well above the region where it can produce an audible peak to colour the reproduction. Though it sounds less bright than a moving-coil unit, it has a more natural and pleasing effect.

The flat response enables maximum gain to be obtained from public address systems before feedback occurs. The ribbon is therefore the ideal choice for this purpose. Not all ribbons have a flat response though, some have a presence peak deliberately introduced by the makers. The response chart should be checked before choosing.

A ribbon microphone should not be positioned too close to the user otherwise the explosive consonants ('p' and 'b') produce a blasting effect; in addition, bass is unnaturally emphasised.

Sensitivity is less than with a moving-coil unit owing to the small area of the ribbon; however, some models incorporate a small, built-in horn in front of the ribbon which increases efficiency. Output from these is not far short of a moving-coil microphone. Natural output impedance is around 0.1 Ω so in all cases an internal transformer is used to provide the required output impedance.

Robustness is not a strong point, and the ribbon can be damaged by shock or by blowing into the microphone. Convenience and reliability are good providing reasonable care is taken in handling.

Crystal/ceramic A slice *of piezoelectric* crystal, secured at one end with a compliant clamp, generates an electrical signal when it is stressed by a vibrating cone in contact with the other. As the natural crystals: Rochelle salt (sodium potassium tartrate), quartz and tourmaline are fragile and adversely affected by high humidity and temperature, more stable synthetic substances such as barium titanate and lead zirconate titanate are used in their place.

High-frequency response is limited to around 10 kHz by the system's inertia, but can be improved at the expense of output by coupling the cone to the crystal through a reduction lever.

Two slices, cemented together so that one is stretched while the other is compressed, form a *bimorph*. Connected in series these give a higher output and cancellation of mechanical-strain non-linearities. Several slices are used to produce a *multimorph* or *sound cell*, in which the sound actuates the slices directly, dispensing with the cone and its resonance.

Output impedance is very high, about 1 MΩ, which gives a high signal voltage to drive valve or f.e.t. input stages, but the low load impedance offered by a bipolar transistor causes severe loss of bass. Another effect of very high impedance is excessive loss of high frequencies over cables of more than a few feet long. Cost is low, but few are used today except for specialised applications such as vibration measurement.

Capacitor A thin, plastic diaphragm coated with aluminium or gold is stretched over a shallow cavity having a flat metal backplate. A capacitance is thereby created which depends on the spacing. It is usually between 20–30 pF, but can be from 5–75 pF. An applied d.c. voltage of from 9–100 V, but usually 48 V, charges the capacitor and electrostatic attraction pulls the diaphragm, keeping it taut.

Movement of the diaphragm varies the capacitance and so a charging current flows in and out of the device. This produces a voltage drop over a load resistor, which must be very high in value if a useful signal voltage is to be developed from the small current flowing. Output impedance is therefore ultra high, around 100 MΩ. A connecting cable of more than a few inches would cause severe loss of high frequencies owing to cable capacitance, so a built-in pre-amp is essential.

The very low mass and inertia of the diaphragm gives a flat and extended response, while output is high because of the built-in pre-amp. Robustness and reliability are reasonable provided high humidity is avoided and the polarising voltage is not excessive. These can produce internal flashovers across the small gap between diaphragm and backplate.

Convenience rating is poor owing to the need for polarising voltage and power for the pre-amp. A supply must thus be provided, along with separate wires in the microphone cable to carry it.

Alternatives are *phantom powering*, whereby the power is conveyed by the audio wires, or a *d.c.-to-d.c. converter* in the microphone to step up the voltage from an internal 1.5 V battery. Current drain is about 5 mA.

R.f. capacitor No polarising voltage is needed for these although a 12 V supply is required. A built-in f.m. oscillator circuit is tuned by the capacitor, thus producing frequency modulations in sympathy with the sound. An integral detector converts these to audio which is fed out via the cable. Apart from a lower supply voltage, the advantage is avoidance of electrostatic tension of the diaphragm and the risk of flashover.

All capacitor microphones are expensive, and this, plus the complications of a d.c. supply restricts their use mainly to recording and broadcast studios where the highest quality is required at any cost.

Electret This is an inexpensive capacitor unit in which a permanent charge equivalent to about 100 V is implanted in the diaphragm during manufacture by heating the plastic between the plates of a charged air-spaced capacitor. The charge does leak away but very slowly; it is claimed to take at least 100 years to reach half its original value.

The diaphragm needs to be thicker than that of the regular capacitor unit so it is less responsive to high frequencies and also has a resonance peak at 5–8 kHz. This is higher than the moving-coil peak; high enough to be flattened with a little treble cut without affecting the upper speech frequencies. Thus the electret is quite suitable for public address use. As performance differs considerably between models, care is needed in making a choice. Most electrets have a falling bass response.

An internal pre-amp is run from a 1.5 V battery which has a low current drain and has a long life. There can be problems with battery contacts and the power switch. Also close miking can cause deterioration owing to condensation. Apart from these snags the low cost makes the electret an attractive alternative to some of the other types.

Power supplies for capacitor units   There are two basic requirements for the 48 V polarising supply: it must be free of mains ripple to avoid hum; and the voltage must remain constant to avoid variations of sensitivity. To fulfil these, a supply with a smoothing resistor and capacitor should be provided with a shunted zener diode to provide stabilising.

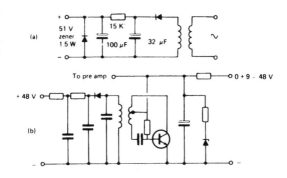

Figure 34. (a) basic supply circuit for capacitor microphones; (b) converter to provide polarising voltage from internal battery.

Phantom powering   To avoid using any extra wires in the connecting cable, phantom powering uses the audio wires and the screen. The positive can be taken to a centre-tap in the mixer input transformer or to the centre of a pair of resistors across the winding.

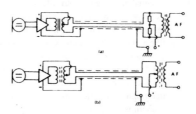

Figure 35. Phantom powering via (a) centre-tapped resistors; (b) transformer.

A–B powering This overcomes the disadvantage of phantom powering which requires the screen to be connected at both ends and thereby forms a complete circuit whereby induced hum and noise could be applied to the input stage. With A–B powering both signal wires are used for the supply and the screen does not form part of the supply or input circuit.

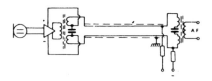

Figure 36. A–B powering. Capacitors between windings prevent a supply short circuit.

Acoustic characteristics

Omni-directional A cone or diaphragm with a sealed back responds to pressure variations on its front. At low and medium frequencies, diffraction occurs so the unit responds to sounds from all directions. Diffraction decreases as the wavelength approaches the diameter of the diaphragm, so the response to off-axis sounds diminishes at high frequencies. The unit thus becomes increasingly directional as the frequency rises, and the smaller it is, the higher the starting point.

With front-propagated waves, direct particle velocity adds to the pressure on the diaphragm, so reducing the omni-directional effect. This is sometimes lessened by fitting a screen of artificial silk between two wire meshes over the front.

Cardioid If the back of the diaphragm is open to the air through side vents, equal pressure on both sides of the diaphragm gives zero movement. Acoustic resistance pads reduce the rear pressure, so creating a *pressure gradient* between front and rear. The diaphragm moves accordingly.

Response to front-propagated waves is greater than to others because particle velocity adds to the force on the front of the diaphragm. The polar response is heart-shaped when plotted on a circular graph, hence the name *cardioid*. The microphones are sometimes termed *velocity* types. Directivity is greater at higher frequencies because high-frequency waves from the front are not diffracted to the side vents and so do not reach the back of the diaphragm.

Change of output with angle of incidence θ is:

$$\frac{1+\cos\theta}{2}$$

Hypercardioid/supercardioid Reducing the acoustic
resistance of the rear vents of a pressure-gradient unit increases
pressure at the back and thereby cancellation of side-
propagated waves. This results in a narrower forward lobe.
However, particle velocity from rear waves can also be
admitted and so produce a small rear lobe. As the prime force
is at the back of the diaphragm, the lobe is negative.

Velocity/ribbon Ribbon transducers are true velocity devices
because back and front of the ribbon is open to the air, so there
is no pressure difference. As the ribbon is actuated only by
particle velocity, it responds to front and rear propagated
sounds but not those coming from the sides. The polar diagram
is thus a figure-eight configuration.

Pads are sometimes inserted behind the ribbon to restrict
rear response. With some models it is reduced to just a small
rear lobe, which makes the overall effect similar to the
hypercardioid.

Gun/interference tube An open-ended tube having a series
of holes or slots along one side is mounted in front of a
cardioid microphone. Front propagated sound waves enter the
end of the tube and also the holes; all reach the diaphragm at

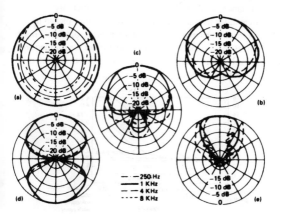

Figure 37. Polar diagrams of various microphone directional
responses: (a) omni-directional; (b) cardioid; (c) hypercardioid;
(d) figure eight/velocity; (c) gun/interference tube.

the same time as all paths are of equal length. When waves come from the side, those which enter the end of the tube follow a longer path than those entering the holes. Thus a delay and phase difference occurs. When the paths are half a wavelength different, cancellation takes place. As there are multiple paths via the line of holes, cancellation occurs over a range of frequencies down to where the tube length equals half a wavelength. Thus high directivity is achieved down to this frequency. Below it, directivity falls off to the point where the unit reverts to a cardioid.

Directivity   Also termed *rejection ratio; discrimination; cancellation; and front-to-back ratio;* this denotes the difference in sensitivity between points of maximum and minimum pickup. Typical values 15–20 dB.

Directivity coefficient/sound power concentration   If an omni-directional and a directional microphone of equal sensitivity are placed in a reverberant field facing a sound source the acoustic power received, hence electrical power produced by the omni, is greater than that of the directional unit, because it receives reflected sound from all directions. The ratio of the two powers is thus a measure of the total directivity of the directional unit and is described by the above terms.

Ambient noise and distance   Because the sound intensity decreases as the square of the distance, a directional microphone can be further away from the source than an omni, by a distance equal to the square root of the directivity coefficient, for the same amount of ambient noise picked up. Thus a cardioid having a coefficient of 3 can be placed √3 or 1.73 times the distance of an omni from the source for the same noise level. Alternatively, the noise will be 1.73 times (4.8 dB) less at the same distance.

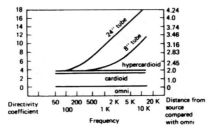

Figure 38. Directivity coefficients and distance from source compared with an omni unit, showing frequency dependence of interference tubes.

Special types of microphone

Multi-pattern microphones   Some capacitor units have two
diaphragms, one either side of the plate, facing to the front and
rear respectively. The basic pattern of each is a cardioid.
Polarising voltage to the rear can be switched off, leaving the
front to serve as a cardioid; or it can be switched on thereby
giving a combined omni pattern; its polarity can be reversed, so
forming positive front and negative rear lobes with cancellation
at the sides, i.e. a figure-eight pattern; or the reverse voltage
can be reduced, giving a small rear lobe and some side
cancellation, thus producing a hypercardioid response.

Zoom microphones   Two cardioid capsules connected in
opposite phase are placed one behind the other to obtain high
directivity by cancellation of side propagated sound. With
frontal sounds the outputs reinforce each other only when
capsule spacing is half wavelength; at frequencies below this
the response falls at 6 dB/octave due to cancellation. To avoid
loss of bass, a frequency-dependent phase-shifter progressively
changes the relative phase until they are in phase at the lowest
frequency. Thus bass response is maintained, although the
directivity decreases to that of a cardioid at low frequencies.
Polar pattern is therefore similar to that of an interference tube
and it can be continuously varied down to a straight cardioid by
controlling the gain of the second capsule to zero. Going
further, the response can be broadened to an omni by fitting a
third cardioid capsule back-to-back with the second. By using
ganged and centre-tapped potentiometers, the gains can be
simultaneously adjusted to give continuous variation from high
directivity to omni with a single control. The device is
designed for use with video cameras to match the sound to the
operation of the zoom lens.

Output variation with angle of incidence θ at maximum
directivity
is:

$$\frac{(1+\cos\theta)}{2}\cos\theta$$

Noise-cancelling microphones   These are usually used for
announcements in areas of very high ambient noise. They
consist of two cardioid or hypercardioid units spaced a few
inches apart and connected in antiphase. Speech is directed at
one of the units but is also picked up by the other, so producing
some cancellation. As the sound pressure is greater at the first
unit, cancellation is only partial. Ambient sound coming from
some distance affects both units equally and so is completely
cancelled.

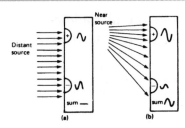

Figure 39. Noise-cancelling microphone using two antiphase units: (a) plane waves from distance cancel; (b) local waves affect one unit more than the other.

Speech wavelengths shorter than the distance between the units are not cancelled at all and cancellation increases with wavelength; so the device has a falling bass response. This, together with the need to speak closely makes the microphone suitable only for the purpose intended.

Parabolic reflectors   Acoustic gain is obtained when a reflector is used to focus sound onto a microphone diaphragm, accurately positioned facing the reflector at its focal point.

Plane waves are increasingly diffracted around the reflector as wavelengths become greater than its diameter. Not being reflected to the microphone, they produce no output and result in a falling bass response.

Waves reflected from the outer areas of the dish arrive at the diaphragm at an oblique angle. At wavelengths shorter than the diaphragm diameter, there is thereby a phase difference across its surface which increases with frequency. Output thus falls at high frequencies.

Overall response is thus a mid-range peak with bass and treble declining rapidly on either side. Its frequency is determined by the reflector size: for a 12 inch diameter, it is about 10 kHz, for 18 inch it is around 7 8 kHz, and for 24 inch approximately 5 kHz. It is also influenced by diaphragm diameter, the smaller it is the higher the frequency.

A cardioid microphone receives no direct sound from the rear, but if an omni is used, some direct as well as reflected sound affects the diaphragm. So when the distance between reflector and diaphragm is equal to a quarter wavelength, the reflected wave is partially cancelled by the direct. This produces a sizeable dip in the response curve.

With smaller reflectors the dip can go below the output level for a microphone without a reflector. However, the dip can be eliminated with a deep reflector which has a short focal length.

The bass response of a reflector with an omni is somewhat better than with a cardioid because it receives direct sound below its bass cut-off point

Polar response of a reflector is highly directional at 5 kHz, but the directivity decreases down to virtually omni-directional characteristics at 250 Hz.

When a cardioid is used, the rear response at 250 Hz is actually greater than that of the forward direction, because the microphone is pointing backward and the reflector is ineffective at that frequency.

Smaller reflectors are less directional than large ones and also have less gain. The poor polar and frequency response at low frequencies make them suitable only for applications involving mostly higher frequency sounds such as bird-song recording.

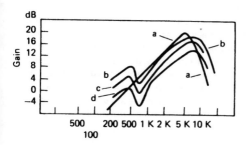

Figure 40. Frequency response of (a) 24 inch unit with cardioid; (b) 24 inch with omni; (c) 18 inch with omni; (d) 12 inch with omni.

Boundary microphones

Boundary microphones are a recent innovation, being produced commercially in the early 1980s. They were developed and manufactured by Crown who used the proprietary name *pressure zone microphone* (PZM). Other manufacturers are now making them.

The operating principle is acoustic, so microphone capsules of varying type and quality may be used by different manufacturers. Caution must thus be exercised. as disappointing results may be due to an inferior capsule or its housing rather than any fault of the principle. Like other types there are good and bad examples.

Undoubtedly the boundary microphone offers many advantages over conventional instruments although it may not be the answer to every microphone problem. Being in its infancy, there is still much to learn as to its use, but the following information is presented as a guide.

Boundary effect   When a sound pressure wave meets a large hard smooth surface such as a wall, it is reflected from it. A microphone placed near such a surface thus receives two pressure waves, one direct, and shortly afterwards, the reflected one. At a frequency having a wavelength of half the distance from the microphone to the surface, and those that are 3 times, 5 times, 7 times that frequency, the reflected wave is in phase with the direct one, so there is a pressure reinforcement and a peak in the microphone response at each of those frequencies.

At the frequency corresponding to a quarter wavelength, and twice, 4 times, 6 times, 8 times that frequency, there is cancellation and a dip in the response. The resulting alternate peaks and dips look like a comb when plotted, hence the term it is given: *comb filter effect*.

In a recording studio where movable baffles are often used to prevent leakage from one sound field to another in multi-microphone set-ups, the vagaries in frequency response resulting from boundary effects can be a major problem. Wherever a microphone is used near a large reflective object such as the back wall of a platform, a speaker's rostrum or even the floor, a comb filter effect is obtained.

If the microphone is moved closer to the surface, the half-wavelength is shortened and the frequency at which the comb filter effect starts, is raised. If it is moved very close, the frequency can be raised above the usable range of the microphone and the comb filter effect disappears.

In addition, at all frequencies, the direct and reflected pressure waves are virtually in phase, hence reinforcement occurs over the whole range. This doubles the overall sensitivity, so giving an acoustic gain of 6 dB.

Pressure area   The distance from the surface at which virtual in-phase reinforcement occurs is a small fraction of the wavelength. So it becomes smaller as the frequency rises, and to obtain the effect at highest frequencies, the distance must be minute.

For the phase difference to produce a loss of only 1 dB, the distance between boundary and microphone diaphragm must be $^1/_{13}$ of the wavelength. At 20 kHz, this is just 0.052 inches. For a drop of 3 dB the distance must be $^1/_8$ of the wavelength which at 20 kHz is 0.085 inches. For a loss of 6 dB the distance is $^1/_6$ wavelength and for 20 kHz this is 0.11 inches.

These losses are subtracted from the 6 dB gain, not from the original microphone sensitivity, and are only for the highest frequencies. At all lower ones the phase difference is less and so also is the loss.

Microphones designed to exploit this effect have their diaphragms facing and mounted very close to a metal plate which functions as the boundary. The area of the plate is extended by mounting it on a table, wall, floor or other large convenient surface which then serves as the main or primary boundary. These instruments are also known as *pressure zone microphones* although this is a proprietary name of one manufacturer.

Off-axis coloration   With a conventional microphone, high frequency sounds coming in off-axis, that is at an angle to the diaphragm arrive at one half of the diaphragm before the other. There is thus a phase difference across the diaphragm resulting in a reduced excursion and lower output. Low frequencies are unaffected. The audible effect is that off-axis sounds are less sharp and clear than those normal to the diaphragm.

With a boundary microphone, the proximity of the plate which blocks and reflects incoming pressure waves, reduces directional effects and phase differences across the diaphragm. The result is that sounds from all angles within the polar response sound the same and there is no coloration of off-axis sounds.

Polar response   The polar response is hemispherical, the microphone picking up sounds from all angles on the microphone side of the mounting surface. The amplitude is uniform over almost the whole hemisphere, dropping about

6 dB only at the extreme edges, at angles from 170–180°. The frequency response over the hemisphere is also very consistent, with little difference between high and low frequencies, a characteristic not obtainable with any other type of microphone.

The polar response can be modified to make it more directional by masking one part of the plate and unit with acoustic foam or carpetting. Alternatively, some models are designed to have a more directional response.

Pick-up of sound from the rear, that is from the opposite side of the boundary surface is dependent on frequency and the size of the boundary. At low frequencies, pressure waves from the rear are diffracted around the boundary and are picked up by the microphone. The device thus has an omni-directional response to these.

Frequencies from the rear having a short wavelength compared to the boundary. are blocked and so are rejected. Rear rejection frequency response thereby depends on boundary size. Given a square panel, for 3 dB rejection, the frequency is $246/d$ in which d is the dimension of one side in feet. The 10 dB rejection frequency is $985/d$. Thus a 2 x 2 ft (0.6 x 0.6 m) panel rejects 500 Hz by 10 dB, and 125 Hz by 3 dB. It also rejects 10 kHz by 20 dB.

Frequency response The front response also depends on the size of the boundary, which is to the microphone what a baffle is to a loudspeaker, though with a somewhat different effect.

Pressure waves arriving from the front are diffracted around the boundary when the wavelength is six times or more the size of one dimension of a square boundary. There is thus no reflected wave so the 6 dB gain disappears. The effect is that of a shelf filter, the gain is flat down to the critical frequency, then falls 6 dB to remain at this lower level until the bass roll-off of the capsule is reached. The frequency is equal to $188/d$, where d is one side of the square in feet.

Reach One of the most remarkable features of this extraordinary type of microphone is its reach. In the near field, the output falls as distance from the source increases in a similar manner to a conventional microphone but in the free field beyond, where there is normally a 6 dB drop for a doubling of distance, the output barely changes.

Boundaries The microphone must be mounted on a large flat reflective surface in order to obtain the rear rejection, bass response, and pressure doubling effects. The effect of size on these functions has already been considered.

For music, the size of the boundary should be about 4 x 4 ft (1.2 x 1.2 m). For speech, a square 2 x 2 ft (0.6 x 0.6 m) should be adequate. Panels this size may seem obtrusive. but clear acrylic plastic can be used, which is visually unobjectionable. If the edges of these catch the light or show up white, they can be painted black to make the panels even less conspicuous.

Diffraction around the edges of the boundary can give rise to comb-filter effects, but these can be minimised by placing the microphone off-centre. A rectangular or asymmetrical shape is better still, just as it is with a loudspeaker baffle. The worst shape is a circle, as then there is an equal path to the edge at all points from the microphone.

According to the application. panels can be supported from the ceiling by piano wire, or made free-standing.

Otherwise, the microphones can be installed in table tops, in lecterns, on walls, or on the floor. In the latter case, it should be mounted on a wooden panel if it is to lie on carpet, as carpet is a poor sound reflector. With a wooden or lino-tiled floor, the microphone can be just placed as it is.

Stereo A pair of boundary microphones can be placed one each on a panel that is spaced from the other and facing the performers. Because of their long reach. these microphones will usually be sufficient to pick up an orchestra or choir. They can be suspended overhead and forward by several feet, and angled downward. In many cases two units placed on the floor will be satisfactory, and this arrangement can reduce audience noise and unwanted ambience.

Two spaced microphones of conventional type have the drawback of poor response to the centre to which both are off-axis. As boundary microphones do not suffer from off-axis coloration, the centre is well reproduced if they are so spaced.

Near coincident stereo A single-unit arrangement is usually provided by a coincident pair of cardioid microphones. Here again the centre is off-axis to both to a certain degree. A pair of boundary microphones can be mounted one on either side of a single boundary plate, which is directed edgeways-on to the centre of the sound source. Each hemisphere picks up its respective channel as well as part of the centre. The closer the assembly is to the performers the wider is the stereo spread.

There is some loss of centre as the response falls to some −6 dB at the edge of the hemisphere. This can be overcome by using two plates in a V formation with a microphone on each and the point of the V directed to the source. The V should form an angle of about 70°. but varying the angle changes the stereo spread. While the reach of boundary microphones permits a location some distance from the performers, this also increases ambience. Distance should thus be adjusted for the desired amount of ambience.

Conferences   A single boundary microphone will pick up around eight persons at a table if installed at the table's centre. For larger numbers more microphones may be required.

An alternative is a microphone placed in the corner between two walls and the ceiling. For this position it will be necessary to remove the plate so that the capsule can be aimed into the corner. A gain of 12 dB is obtainable and remarkable clarity. For large conference rooms, one microphone in each corner may be required.

Stage   Three boundary microphones placed at the footlights across the stage will pick up most of the action. If the system is stereo, the centre microphone should be replaced by a V-shaped stereo pair. This avoids anomalies for members of the audience seated on the extreme sides.

For a deep stage, other boundary microphones may be installed at any convenient position toward the rear. Scenery has proved an ideal boundary.

Opera   A similar arrangement as for the theatre stage, but an additional pair will be needed in the orchestra pit. These should be spaced apart on the wall behind the conductor, facing the orchestra.

Choir   A single boundary microphone on an acrylic panel suspended over and to the front of the choir, angled downward, will pick up all sections even the back row, due to the long reach of the device. Conventional microphones usually fade the further sections. For stereo, a spaced pair, single panel bi-polar or V panels can be used. Floor mounting is another possibility, especially if the choir are arranged in tiers.

Public speaking   A single boundary microphone can be fitted to the rostrum or lectern. Some models can be flush-mounted to give a low profile. Notes or papers laid over the microphone do not affect its performance providing they are not too thick. Alternatively the microphone can be mounted on the floor a little to one side of the rostrum so that it is not masked by it. If the floor is carpeted a wooden panel should be laid beneath the unit.

Feedback   Because of the hemispherical polar pattern, a floor mounted unit may pick up sound from the loudspeakers and give early feedback. Carpeting folded over the back of the microphone will make it more directional. A vertical panel or wall mounting gives better results.

Microphones in the sound field

Distance effect As most microphones respond to pressure waves, their output decreases with distance in accord with sound pressure. Thus for a point source in a far field, doubling the distance from the source results in a 6 dB drop or half the output. In a diffuse field the output may vary little with distance.

Proximity effect A velocity or pressure gradient microphone responds in part to particle velocity. This decreases according to the square of the distance in the near field, but in proportion to distance in the far field. The division between near and far fields depends on frequency: the near field is large for low, but small for high frequencies. So when a pressure gradient microphone is used close to the source, it is in the near field for low frequencies, but in the far field for high ones. Thus the low frequencies are emphasised with close use.

Proximity effect is obtained with figure-eight, cardioid and hypercardioid units, but not omnis. For speech it is a disadvantage except with lapel microphones which normally are deficient in bass.

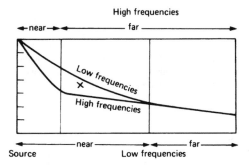

Figure 41. Near and far field response for pressure gradient units at low and high frequencies. At distance X, there is less attenuation of low than high frequencies because it is in the low-frequency near field.

Pressure build-up With a cone or diaphragm, pressure waves shorter than its diameter coming from the front are reflected back. These build up a back pressure which produces a broad peak in the high frequency response on axis. Off-axis response is normal. The effect is not obtained with ribbon transducers.

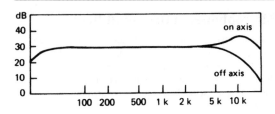

Figure 42. A typical response at 0° and 90° incidence.

Direct/diffuse fields The distance at which a microphone is placed from the sound source affects not only the output and to some extent the frequency response, but also the proportion of direct to diffuse sound it receives. Direct sound decreases with distance, but diffuse sound resulting from multiple reflections is fairly constant.

So the greater the distance from the source, the more in proportion will be the diffuse sound. The result for a too distant microphone can thus be hollow and reverberant. If unavoidable, such as when filming indoor locations, the diffuse sound can be reduced by using a hypercardioid or interference tube, and by acoustically deadening the environment with suitable absorbents. The problem does not exist outdoors or in large studios with built sets, but similar precautions may be required to reduce extraneous noise.

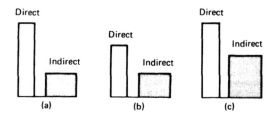

Figure 43. (a) ratio of direct-to-diffuse sound with close microphone position; (b) ratio for more distant microphone; (c) amplifying direct sound to that of the close position also increases diffuse sound.

Multiple sources   Where a single microphone is required to pick up a multiple source such as a choir or orchestra, it should be placed at such a distance and position that the paths from each source are as equal as possible. Otherwise, the nearest ones will dominate. As this is best achieved by a distant position which also produces excessive diffuse sound and low output, a compromise is usually necessary.

Stereo   With a crossed pair of directional microphones each picks up direct sound from one side only. The diaphragms are spaced at about the same distance as the human ears. Figure-eights capture the full ambience of the hall, or it can be reduced if desired by using hypercardioids.

A dummy head with microphones in the ear positions is used to give *binaural* reproduction, while with *holophony*, the convolutions and internal passages of the ear are duplicated to simulate the reflections and directional phase coding obtained in human hearing (see pages 261–3).

An alternative is to space two omnis well apart in the positions that the reproducing loudspeakers would occupy. Cardioids can be used to reduce excessive reverberation. With either method, *spot microphones* are added to pick up weaker individual instruments. Their outputs are sometimes delayed to prevent them sounding too 'forward' in the stereo image.

Wind and popping   Rushing noises are produced by wind when microphones are used outdoors. Plastic foam shields can reduce this by up to 20 dB. Better reduction can be achieved with large elaborate shields of layered close wire mesh and silk. 'Popping', the blasting noise made by the consonants 'p' and 'b' when the unit is used too close to the mouth, can be reduced by the same means. Pressure gradient models are more prone to this effect than omnis.

Microphone electrical parameters

Sensitivity   Various methods of specifying sensitivity are used. All relate electrical output to a given sound pressure. When measured in a free field without a load the terms *free field sensitivity* or *response coefficient* are used. The unit is the mV per Pascal, or its equivalent the mV per Newton/m^2. This replaces the mV per μbar and the mV per dyne/cm^2, used formerly.

Sometimes the output is expressed in minus dB. This has reference to 1 V, so −60 dB = 1 mV. It should be qualified by the suffix V, thus dBV or the description (0 dB = 1 V) but these are often omitted.

Off-load signal voltage varies according to the impedance of the circuit, so is not an informative measure of sensitivity unless the impedance is known. Comparisons can thus only be made between models of the same impedance.

To take account of impedance, the milliwatt is often used as a reference into a standard 600 ohms. The description is (0 dB = 1 mW into 600 Ω), and the term is dBm. The voltage at 1 mW is 773 mV. These are different ways of expressing the output of the same microphone:

obsolete units	0.1 mV/μbar
	0.1 mV/dyne/cm^2
	−80 dBV/μbar (0 dB = 1 V)
	−58 dBm/10 dyne/cm^2 (0 dB = 1 mW into 600Ω)
current units	1.0 mV/N/m^2
	1 0 mV/Pa
	−60 dBV/Pa (0 dB = 1 V)
	−58 dBm/Pa (0 dB = 1 mW into 600Ω)

Impedance   The capacitive reactance of screened cable has a more significant effect when shunted across a high impedance circuit than a low. There is thus a greater loss of high frequencies at high impedance when there are long cables.

Valve equipment has a high input impedance and requires a high signal voltage to drive it. This can be provided only by high impedance microphones of 47 kΩ or more. If long cables are used, low or medium impedance models must be employed with step-up input transformers.

Transistor input stages are of much lower impedance, but low impedance microphones of 30–50Ω give insufficient signal voltage and so result in a poor signal/noise ratio. Medium impedance, 200Ω–1 kΩ give adequate signal level with little loss of h.f. over moderate cable lengths. Most microphones are now made within this range of impedances.

Input impedance   The source impedance of the microphone is effectively in series with the generator. It thus forms the top leg of a potential divider of which the amplifier input impedance is the bottom. When the two are equal, the signal voltage across the input is half the total. So to avoid signal loss, the input impedance should be 5–10 times greater than the microphone impedance. A low value load though can dampen peaks in moving coil units if the output is sufficient to offset the loss.

Effect of cable capacitance   Typical capacitance of microphone cable is 200 pF per metre, which is shunted across the amplifier input impedance. This chart plots the reactance of lengths up to 50 m at 6 frequencies from 500 Hz to 20 kHz. The shunting effect of a length of cable at various frequencies can thus be determined and its effect calculated.

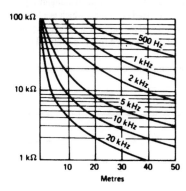

Figure 44. Reactances of 200 pF/m screened cable at various lengths and frequencies.

Frequency response   Published response charts are averaged over a number of units. Better instruments come with individual charts. Pressure gradient units often show several bass curves for different source distances to account for proximity effect. Some models have a switch to curtail the bass for speech. Two treble curves giving a different on and off axis response due to pressure build-up, are shown in some charts.

In all cases care must be taken to observe vertical calibration; curves having 4 dB units look a lot smoother than those with 1 dB divisions.

Noise Thermal agitation in conductors and fluctuations in pre-amp current produce self-generated noise in microphones. There are several ways of expressing this. One is its measured value in μV. As some noise frequencies are more obtrusive than others, the value can be measured after filtering according to a weighting curve such as the DIN 45–405. It is then termed the *weighted noise* figure.

This voltage can be expressed as if it had been generated by an external sound. It is then termed *equivalent noise*, and like other sounds is given a dB value related to the hearing threshold of 20 μPa. The formula is:

$$20\log\left(\frac{V_n \times 10^6}{V_s \times 20}\right)$$

where V_n is the noise voltage; and V_s the rated sensitivity both in μV.

Another method of specifying noise is as a *signal/noise ratio*. Unlike an amplifier there is no maximum signal to use as a reference, so it is fixed at 1 Pascal which is a SPL of 94 dB. Conversion of unweighted equivalent noise to signal/noise ratio or vice-versa can thus be simply performed by subtracting from 94.

Typical weighted noise values: capacitor, 3.6 μV; moving coil, 0.26 μV.

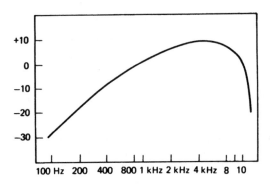

Figure 45. DIN 45–405 Noise-weighting curve.

Hum Dynamic microphones are prone to pick up hum from the magnetic fields of electrical equipment. A maximum sensitivity figure is sometimes quoted. The unit is the μV per μTesla. Standard specified density is 5 μT.

Overload Capacitor microphones can be overloaded at high sound levels so maximum levels are often specified. Typical levels are 20–30 Pa (120–124 dB). Some low output models accept up to 50 and even 100 Pa (128–134 dB), while one industrial instrument with very low output will measure up to 500 Pa (148 dB). Dynamic (moving coil) microphones cannot be overloaded under normal conditions and so no figure is given for them.

Distortion Total harmonic distortion (THD) specifications consist of the sound pressure level that produces a given amount of distortion. The standard distortion level is 0.5°%, but cheaper models often use 1.0%. A typical specification is: SPL for 0.5% THD = 30 Pa.

Microphone input circuits For short cable lengths in non-professional applications, the *unbalanced* circuit is usual. The cable has a braided or lapped screen which serves as a earthed return path as well as shielding the single 'live' centre conductor from hum fields. Being part of the input circuit, the screen can add hum or noise voltages induced into it, to the low level signal.

The *balanced circuit* avoids this by isolating the screen at the microphone end and using twin conductors for the signal. These are balanced about earth by a centre-tapped input transformer or a suitable transistor circuit. They are also twisted together so that any penetrating magnetic field affects both equally and is cancelled out by the balancing circuit.

A *quasi-balanced circuit* offers some of the benefits of the balanced one with an unbalanced input socket. Twin screened cable is used with the screen isolated at the microphone. The two conductors convey the signal, but one is earthed with the screen at the input socket.

Microphone transformers Owing to the very low signal levels, input transformers are extremely vulnerable to hum. They can be eliminated with modern semiconductor circuits, but if used should always be screened with mumetal and rotated around single-point fixing for minimum hum level.

Microphone applications

Application	Transducer					Polar response					
	Capacitor	Electret	Moving coil	Ribbon	Crystal	Omnidirectional	Cardioid	Hypercardioid	Figure-eight	Gun	Reflector
Announcements		•	•				•	•			
Bass drum			•				•				
Bassoon	•		•	•			•				
Birdsong		•	•								•
Brass band				•			•		•		
Brass solo			•				•				
Cabaret		•	•				•	•			
Cello	•		•	•			•		•		
Choir (mono)	•			•		•	•			•	
Choir (stereo)	•			•			•	•		•	
Clubs		•	•				•	•			
Conferences		•	•				•				
Cor anglais; oboe	•	•		•			•				
Demonstrations		•	•				•				
Dinners		•	•	•		•	•				
Filming	•		•							•	
Flute; clarinet	•	•	•	•			•				
Interviews			•			•	•	•			
Lectures		•					•				
Orchestra (mono)	•		•	•		•	•		•		
Orchestra (stereo)	•			•			•	•	•		
PA indoors		•	•				•				
PA outdoors			•				•	•			
Paging		•	•				•				
Percussion	•	•		•			•				
Piano solo	•		•	•		•					
Piano ensemble	•		•	•			•				
Pop groups			•						•		
Radio transmitting		•	•			•	•				
Reporting	•		•					•			•
Saxophone			•	•			•				
Singers, female	•	•		•			•			•	
Singers, male	•			•			•			•	
Stage plays		•	•			•	•				
Strings	•	•		•			•				
Studio speech	•			•		•	•		•		
Tape recorders		•	•			•	•				
TV	•		•					•		•	
Valve amplifiers		•	•	•	•	•	•				
Wild life	•		•					•		•	•

The most suitable types for specific jobs are shown. Where more than one is indicated, particular conditions may dictate which is preferable.

Microphone connectors Several types of connectors are in general use; some are used on the microphone and on the mixer input, others are used solely for the mixer.

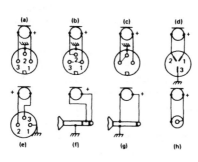

Figure 46. (a) DIN balanced; (b) DIN unbalanced; (c) alternative DIN unbalanced; (d) Tuchel balanced; (e) XLR balanced; (f) jack balanced; (g) jack unbalanced; (h) phono unbalanced, domestic equipment only. (a–e viewed from free end of microphone pins, and solder-tag end of plugs).

The gramophone

Studio techniques

Acoustics   The correct reverberation characteristic for the music is obtained by positioning of movable absorption panels in the studio.

Microphone techniques   These vary, but most use a main stereo pair with *spot microphones* to cover weaker instruments. Delay lines may be used to synchronise these with the signal from the further main microphones.

Pan-potting   The output from the spot microphones is mixed into the two stereo channels in proportion to their position. This produces amplitude but not phase differences. The stereo image of these is thus imperfect.

Depth   More distant instruments produce a greater amount of reverberation in proportion to direct sound than closer ones, thereby giving the illusion of depth. Near sound sources propagate spherical wavefronts whereas distant ones have a nearly plane front. So for the same angle of incidence there is a greater phase difference across a stereo microphone pair for a near than a far sound. This adds to the effect of depth.

Master tape   A tape having from 16 to 32 tracks records all microphone outputs separately so that they can be mixed down to the stereo pair later. Uncorrectable balancing errors at the recording session are thus eliminated. Some tracks are often combined and re-recorded on a spare track for further mixing and re-recording. Thus some material may be re-recorded three or four times.

Noise reduction   Each recording generates noise; the first re-recording has 6 dB more noise than the original, the second 3–5 dB more than the first, and so on until some 12 dB extra noise has built up. *Dolby A* noise reduction reduces this by 10 dB. It splits the signal into two paths, the *main* and the *sidechain* in which it is divided into four parallel paths, each responding to a specific frequency band. These are: low–80 Hz; 80 Hz–3 kHz; 3 kHz–high; 9 kHz–high. The signals in each are progressively compressed above a 40 dB threshold thus emphasising those below it. These are then added to the main path and recorded. After replay, the sidechain frequencies are subtracted so restoring the original balance while reducing low-level recording noise. The levels to record and playback processors must be identical, so a 700 Hz reference tone is recorded at the start of a tape to establish the correct playback level.

Compression   A *compressor* avoids overloading the tape on loud signals. A *threshold control* sets the starting level below which the signal is unaffected. A *slope control* selects the degree of compression. Thus a 2:1 ratio reduces signals above the threshold by a half; a 4:1 by a quarter.

Limiting   All signals above a set level are reduced to that level. Unlike the compressor there is no slope or gradual attenuation. A somewhat higher recording level can therefore be used without fear of a peak overloading the tape. Limiters and compressors are often used together.

Expander   Does the opposite of a compressor; it expands signals above a set threshold. If sounds above a threshold are expanded, it is equivalent to attenuating those below it. It can thereby reduce noise or low-level background sounds and when so used is called a *noise gate*. By attenuating decaying sound, reverberation time can be reduced.

In addition to these methods of recording natural sound, there are many devices used in rock music which introduce distortion for special effect. Here are a few examples.

Artificial reverberation   Can be used to make mediocre vocalists sound better, but also applied in varying degrees to instrumental sounds. Short single delays added to the original give *overdubbing* effects, sounding like two voices or instruments in unison. Several short delays can make two or three voices sound like a chorus. Longer delays give echoes.

Echo plates and spiral steel springs having transducers to impart and pick up sound vibrations have been the standard means of obtaining these effects. Now electronic *bucket-brigade* delay lines are supplanting them. These consist of up to several thousand electronic capacitors connected in line in an integrated circuit. The signal is sampled at three times its highest frequency, and a charge corresponding to these samples is passed down the line. Their rate of progress is determined by pulses generated by an external clock. Thus the total delay can be altered by changing the clock frequency and multiple delays are obtained by tapping off at various points along the line.

A mixture of various types of delay and reverberation applied at different points and to different instruments can often be found on the same record.

Fuzz   The deliberate overloading of an amplifier to produce clipping of peak waveforms. This is more usually done by the performer.

Musical additions   A series of arpeggios can be generated from single notes, and various rhythm beats can be synthesised.

Phasing/flanging Originally achieved by running the same signal on two recorders, one of which was intermittently slowed by an obstruction held against the tape spool. When combined there were alternate cancellations and reinforcements of the two signals. This was termed *phasing;* an electronic version which did not have quite the same effect was called *flanging* (In American terminology, the appellations are reversed.)

Pitch changing Pitch can be increased or decreased without altering the timing. One stereo channel can be gradually flattened while the other is made sharp; and the effect can then be changed over. Pitch changes can be made over a range of up to two octaves, so a unison harmony part of any desired musical interval can be generated from the original.

Pitch changing is often used to help vocalists hit the high notes they cannot reach or extend the bass below the normal instrument range.

Speed changing Speed can be changed without varying the pitch. Instrumentalists who cannot play fast enough can thereby be speeded up. Sometimes different tracks are combined at different speeds for effect.

Sustaining A single note can be sustained indefinitely by 100% feedback, and modified as it plays. This is termed *spin.*

Vibrato/tremolo Rapid minor variations of pitch as performed by string players is termed vibrato, while variations of amplitude usually produced by wind instrument performers are called tremolo. Both of these can be electronically added to any instrument or vocal passage.

The record groove

The two walls of a stereo disc are cut at 45° angles from the perpendicular. Each recedes or advances independently of the other in sympathy with the recorded sound, the outer carrying the right-hand and the inner carrying the left-hand signal.

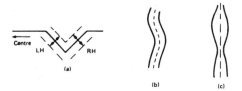

Figure 47. Stereo record groove: (a) 45°/45° angles; (b) signals in phase in both channels, lateral groove movement; (c) signals out of phase, vertical groove displacement.

When the signal in both channels is equal and in phase, the groove moves laterally, but when they are out of phase it moves vertically. Two generators in the pickup are so mounted as to respond only to one plane of motion of the stylus. Groove width thus varies, but it should not be less than a specified 1 mil (25 µm). In practice 2 mil is the observed minimum to avoid groove-jumping with cheaper equipment.

The cutting stylus, which is a triangular chisel shape, was inclined from the vertical by 15° from 1964–1972, and 20° since then according to the IEC standard. However. a +5° tolerance is permitted. Owing to the elasticity of the disc material, there is a springback effect after the cutter has passed. This reduces the tracking angle which together with the tolerance means that the angle can differ considerably from one record to another. The cutter traverses the disc tangentially, in a straight line from disc edge to centre.

The cutter response is termed constant velocity, which means its velocity is the same for all frequencies. The amplitude therefore increases as the frequency drops, at the rate of 6 dB/octave. It would thus be some 16 times greater at 30 Hz than it is at 15 kHz.

Large low-frequency stylus excursions are avoided by cutting the bass while treble is boosted to improve signal/noise ratio. These contours roll off either side of a short flat region centred on 1 kHz, to form the *RIAA recording characteristic*. These are the specific levels:

Frequency	dB	Frequency	dB
1000	0	1000	0
700	−1.2	2000	+2.6
500	−2.6	3000	+4.7
400	−3.8	4000	+6.6
300	−5.6	5000	+8.2
200	−8.2	6000	+9.6
100	−13.1	8000	+11.9
50	−17.0	10000	+13.7
30	−18.6	12000	+15.3
		15000	+17.2
		20000	+19.6

Disc manufacture

Disc cutting The signal from the tape is fed to the disc cutter and modified to the RIAA recording characteristic. It is presampled 1.1 seconds ahead of the main playback head to provide a control signal for the cutter-head radial drive, or the main cutter signal may be delayed. By this means groove spacing can be varied according to the amplitude of the signal, thereby achieving maximum playing time. Spacing can vary from 130 to 390 grooves per inch. A heated triangular chisel, actuated by two helium-cooled coils. cuts the groove in a lacquer-coated aluminium disc 14 inches in diameter for a 12 inch record and 10 inches for a 7 inch.

Master disc When completed, the lacquer disc is coated with silver to make it conductive for electroplating with nickel. This is done in two stages: a slow, precise one to plate the grooves with 1 mil of nickel, followed by a faster build-up of 20–30 mils. The nickel is then parted from the lacquer to form a negative metal master disc having ridges in place of grooves. Although this could be used to stamp out the final pressings, it is much too fragile.

Mothers Impressions are made from the master by a similar plating process. Up to four mother discs can be produced, but for top quality only two are made. These are positive so cannot be used to make pressings, but they can be played.

Stampers From each mother up to six negative stampers are produced. These are chromium-plated to make them harder and to give the pressings a shiny surface.

Pressings Two stampers, one for each side of the disc, are steam heated in the press as the pressure is applied to the plastic material, which is then cooled before removal. Too short a heating/cooling cycle can cause minute discontinuities and pits. On removal, the flash is trimmed from the pressing which is then labelled and sleeved.

Testing A few initial pressings are first made and sound checked. Faults common to all must be due to an earlier process. Only when all are traced and rectified is production started. Each disc is visually inspected after pressing, but some clicks and pops can get by.

Disc materials   The main materials used are a vinyl copolymer, a stabiliser, a lubricant, and a pigment, usually carbon black. The latter produces the traditional black colour, but other colours including white have been used. These are compounded and extruded or fed into the press in the form of small cakes.

Durability   Disc material is subject to two types of deformation during playing: elastic, from which it returns, hence is temporary; and plastic, which is permanent, hence is termed 'wear'.

Direct disc   The dynamic range available on disc is limited mainly by the master-tape recording. To extend it, some recordings have been made directly on to the master disc as were the old 78 r.p.m. records. Mothers, stampers and pressings are then made from it in the usual way. The whole of one side of an LP must be recorded without fault in one take, a difficult undertaking for the performers. Many lacquers are scrapped before a satisfactory master is produced.

Reproducing styli

Spherical stylus   The most common stylus for low-cost players is cone-shaped with a spherical point having a radius of 0.6–0.7 mil (15–18 microns). While placing the stylus at an optimum position halfway up the groove walls, its contact span is too large to follow short wavelengths. When the span equals a half-wavelength it will trace an amplitude of up to 0.2 of that half-wavelength.

The contact span is 1.4 times the tip radius, so for a 0.7 mil radius it is 0.98 mil. Frequencies corresponding to this half-wavelength are 9.8 kHz for the start, and 48 kHz for the end of the disc, at which amplitudes can be traced up to 0.98 x 0.2 = 0.196 mil. Higher frequencies can be traced but at lower amplitudes.

As the disc cutter is triangular, the groove width is narrower at the sides of the recorded wave than at the top and bottom. Thus the stylus is squeezed into spurious vertical motion twice for each cycle, so producing second harmonic distortion. An average of 6.4% is produced at 8 kHz with 5 cm/sec recorded velocity. This is called *pinch effect*.

Bi-radial/elliptical stylus   The major radius which sits across the groove is similar to that of the spherical stylus and so it maintains the correct position on the groove walls; but the minor radius can trace shorter wavelengths.

Common sizes of bi-radial styli

mil	μm	mil	μm	mil	μm
0.7x0.2	18x5	0.6x0.3	15x8	0.9x0.2	23x5
0.7x0.3	18x8	0.8x0.2	20x5	0.9x0.3	23x8
0.7x0.4	18x10	0.8x0.3	20x8	0.9x0.4	23x10

For a 0.3 mil minor radius the contact span is 0.42 mil. Corresponding half-wavelength frequencies are 23.6 Hz for the start and 11.4 Hz for the end of the disc, at which it will trace amplitudes up to 0.084 mil. As the styli are closer to the shape of the cutter there is less pinch effect. Second harmonic distortion is thereby lower, averaging 4.0% at 8 kHz with 5 cm/s recorded velocity.

While pinch effect is less and high frequency tracing is better with a smaller minor radius, the area of contact is also smaller, and this puts greater pressure on the groove wall for a given tracking force. Small radii should thus only be used with cartridges having low tracking force.

Shibata/Ichikawa styli   These use a parabola instead of a cone shape and so achieve a contact area four times greater than that of the spherical stylus. This improves tracking, reduces wear and doubles the stylus resonant frequency.

Pranmanik stylus   Adequate contact area is maintained with a small minor radius. Contact area radius increases from the tip to the shank.

Van den Hul stylus   Wedge-shaped from the side, but mitre-shaped from the front. The almost straight sides achieve contact over most of the wall surface, with a contact area radius of only 0.14 mil (3.5 μm) giving a theoretical 70 kHz tracing ability at the disc centre.

Materials   Sapphire has given way to diamond tips mounted in metal shanks welded to the cantilever. Acceleration of up to 1000 gravities requires a very low stylus mass. So as metal adds mass, some are made of pure diamond mounted directly, and are called *naked diamond* styli.

Cantilevers   These can contribute up to 60% of the effective mass, so materials must have a low mass yet be very rigid. Some of these in order of merit are: tubular aluminium, beryllium, sapphire, boron. Owing to the leverage ratio, a short cantilever transmits a large movement to the generators and vice-versa, but the ratio reflects the mechanical effect of the load. So small cantilevers give a large output but also have a large effective mass.

Figure 48. Pinch effect. Recording cutter in various positions along the groove showing how groove width depends on the angle of the groove to the cutter. A spherical stylus is 'pinched' in the narrow portions so giving spurious vertical movement. An elliptical stylus is less affected being closer to the shape of the cutter.

Figure 49. Short recorded wavelengths: (a) a spherical stylus has too large a contact span to trace the shortest wavelengths; (b) an elliptical stylus can trace them with its minor radius.

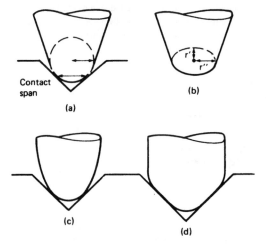

Figure 50. Some stylus contours: (a) spherical stylus; (b) elliptical; (c) Shibata; (d) Van den Hul.

Stylus tracking

For distortionless translation of the groove modulations into electrical signals, the reproducing stylus should follow exactly the same path at the same angle as the recording cutter. A number of factors make this ideal unattainable in practice.

Stylus skew   If the major axis of an elliptical stylus is not in perfect line with the disc radius, one side will be slightly ahead of the other in the groove. Spurious phase differences between channels will thus result, so degrading the stereo image. These are not produced by a spherical stylus. Skew is caused by incorrect mounting of the stylus on its cantilever (manufacturing tolerance is ±3°), an off-centre cantilever. and lateral tracking error.

Vertical tracking error   In following the vertical groove modulations the stylus describes an arc about a fulcrum formed by the anchored end of the cantilever. To clear the disc surface the fulcrum must be raised above it, hence the arc is tilted forward. This is known as the *tracking angle*, and should correspond with the angle of inclination of the recording cutter, termed the *slant angle*.

Deviations from the now standard $20°$ recorded slant-angle are common due to manufacturing tolerances of ±5% and lacquer springback (the partial restoration of the disc lacquer contour after the cutter has passed, due to its elasticity). Cartridge tracking angles are not always accurate. The result is *vertical tracking error* which produces mainly second harmonic and intermodulation distortion.

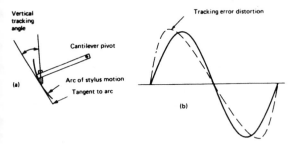

Figure 51. (a) vertical tracking angle; (b) distortion of sine wave cause by both vertical tracking error and lateral tracking error.

Lateral tracking error   The recording cutter head forms a
tangent with all parts of the record groove by moving across
the disc in a straight line, whereas the reproducing pickup
traverses it in an arc.

The axis of motion of the stylus is thus different to that of
the cutter over a large portion of the disc surface. This is
lateral tracking error, which produces second harmonic and
intermodulation distortion.

Second harmonic distortion   The second harmonic
distortion can be calculated from:

$$\delta = \frac{28.73\phi V}{r}$$

in which δ is percentage second harmonic distortion; ϕ is
tracking error in degrees; V is the peak recorded velocity in
cm/s; and r is the radius from the centre of the disc in cm.

A straight arm forms a tangent at one point in the groove,
but if an offset is formed by bending it near the cartridge, two
tangential points can be obtained at which tracking error is
zero. Some designs equally space the zero points from the start
and finish of the groove to equalise maximum tracking error
either side of and in between the two points. However,
distortion for a given tracking error increases as the groove
velocity decreases, and so is at a maximum near the centre of
the disc. It is better, therefore, to arrange the second point to be
nearer the disc centre so that the distortion is equalised rather
than the tracking error. The optimum distances from the centre
for zero tracking distortion are:

$$d_1 = \frac{2r_1 r_2}{\left(1 + 1 \div \sqrt{2}\right)r_1 + \left(1 - 1 \div \sqrt{2}\right)r_2}$$

and

$$d_1 = \frac{2r_1 r_2}{\left(1 - 1 \div \sqrt{2}\right)r_1 + \left(1 + 1 \div \sqrt{2}\right)r_2}$$

in which d_1 and d_2 are the distances, and r_1 and r_2 are the
respective radii from the centre to the finish and start of the
modulated groove. For a 12 inch disc which starts at 145 mm
(r_1), and finishes at 60 mm (r_2), the optimum distances are
120 mm and 65.6 mm.

The effective length of the arm L is the straight distance from the pivot to the stylus, while D is the distance from the pivot to the turntable centre. The difference between them is the *overhang*. Optimum value of D:

$$D = \sqrt{L^2 - d_1 d_2}$$

where d_1 and d_2 are the optimum zero points. For a 12 inch disc these are 120 mm and 65.6 mm, so the pivot to turntable-centre distance is:

$$D = \sqrt{L^2 - 7872} \text{ mm}$$

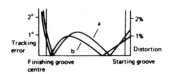

Figure 52. (a) optimum points for minimum tracking error; optimum points for minimum distortion, achieving below 1% second harmonic.

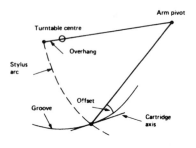

Figure 53. Arm geometry showing offset and overhang.

Offset Optimum offset angle using 120 mm and 65.6 mm zero points calculated from arm length. For a 230 mm (9 inch) arm it is 23°.

$$\phi = \frac{5315}{L} \text{ degrees}$$

Pickup cartridges

There is a variety of ways whereby the motion of the stylus can be translated into an electrical signal. Most of these are grouped into the class known as *magnetic*, because they produce relative movement between a coil and magnetic field.

They are *velocity* devices because the magnitude of the induced signal depends on the velocity of the stylus movement rather than its amplitude. But as tracing speed increases with amplitude so does the output signal. However, stylus velocity also increases with frequency, giving a rising treble response. This is the opposite of the recording cutter characteristic, but as that was modified by the RIAA curve, so also inversely must the response of the magnetic cartridge.

At present the moving-magnet and moving-coil types are most in favour, but other types as well as non-magnetic varieties are here described.

Moving-iron   An armature actuated by the stylus, passes through the centre of a coil which is enclosed between the poles of a permanent magnet. Movement of the armature disturbs the magnetic field through the coil. All the early mono pickups used this principle.

Moving-coil   A pair of tiny coils are mounted so that they can rock about an axis between the pole pieces of a magnet. Each responds only to one plane of stylus movement. To minimise mass, the coils have few turns and so are low impedance and give low output; they thus need special pre-amps or transformers. Some have been produced using many turns of very fine wire, thereby giving higher impedance and output.

Moving-magnet   The stylus cantilever is joined to the free end of a lightweight cylindrical magnet made of material such as samarium cobalt. It is pivoted at the far end, and surrounded by four yoke pieces spaced at $90°$ intervals. Each opposite pair of yokes forms a magnetic circuit through one of two coils. Movement of the magnet in one direction induces a field in the appropriate pair of yokes and from thence to the coil.

Induced-magnet   The principle is similar to that of the moving-magnet except that a stationary magnet induces a field into an armature which is moved by the cantilever. It was a useful alternative when a magnet could add appreciable mass to the cantilever. Lightweight magnetic materials have now overcome this snag, so the induced magnet is less attractive.

Variable-reluctance   Reluctance here is the resistance to a magnetic field offered by an air gap. One pole piece of a magnet is split through a pair of coils to twin faces, in proximity to which is a ferrous moving member linked to the stylus and connected to the other pole. Movement varies the air gap between the member and one face, hence also the magnetic flux through the circuit and coil.

Ceramic   Two slices of piezo-electric materials such as barium titinate or lead zirconate titinate are set at angles, and stressed by stylus movement so producing a voltage. Output is proportional to the amplitude rather than velocity of the recording signal and is thus the approximate inverse of the RIAA recording characteristic. No equalisation is therefore required. A load of not less than 1 MΩ is needed, lower values causing loss of bass. If used into a magnetic input socket with a 47 kΩ load, the RIAA equalisation will roughly compensate for the bass loss.

Cartridge parameters

Effective stylus mass   This is a sum of the actual mass of the stylus, the cantilever and that of the cartridge's moving parts divided by the leverage ratio of the cantilever. Approximate proportions of each can be: 10%, 60%, and 30% respectively. This mass must be accelerated by the record groove to values up to 1000 g at which the vinyl can flow as liquid. Permanent groove deformation is proportional to mass and inversely proportional to stylus contact area.

   The *stylus resonance* which should be above the cartridge frequency range, is governed by its mass reacting with the groove elasticity. The lower the mass the higher the resonance. These are some approximate values:

4 mg	1 mg	0.5 mg	0.25 mg
10 kHz	20 kHz	30 kHz	40 kHz

Most quality cartridges have a lower stylus mass than 0.5 mg, but under 0.15 mg has been achieved.

Compliance   The opposite of stiffness, the unit was the 10^{-6} cm/dyne, but is now 10^{-3} m/N, which is the same value. Compliance depends on the elasticity of the stylus damping which varies with temperature. Butyl-rubber, used in most cartridges, can increase compliance by up to 50% for a temperature rise from 68°F–95°F, and decrease it by 50% for a temperature drop from 68°F–40°F. Other materials have been developed that are less temperature conscious.

The low-frequency arm resonance is determined by its mass reacting with the stylus compliance. High compliance needs a low arm mass to place the resonance at the optimum frequency. At one time high compliance was considered a 'goodness factor' for a cartridge, but too high a compliance is undesirable, and it is far less important than stylus mass.

Playing/tracking/force/weight   Violent oscillations of the groove walls tend to throw the stylus out, so downward force is required to retain it. This must overcome the maximum acceleration (1000 g) of the stylus mass multiplied by a factor of 1.4 due to the 45° groove wall angle. So for a mass of 0.5 mg, a force of 0.5 x 1000 x 1.4 = 0.7 g is needed to equal the stylus force. As this must be overcome with a safety margin, plus allowance for mechanical resistance, arm friction and side-thrust, it must be multiplied further. A factor of x2 would be appropriate.

Excessive weight causes wear on the groove walls, but insufficient weight is worse as the stylus can lose contact then scrape the wall when it re-connects. Weight should therefore be adjusted toward the upper limit of the maker's specified range, but should not be exceeded.

Output voltage   Usually quoted for a recorded velocity of 5 cm/s at 1 kHz.

Channel separation/crosstalk   Signal from one channel appears in the other owing to mechanical and inductive coupling. The figure indicates the amount of attenuation in the opposite channel at 1 kHz. Separation is about 6 dB less at 10 kHz, and falls also in the bass. Poor separation blurs the stereo image. Normal range is from 20–35 dB.

Channel balance   The output difference between the two channels for the same input. Unbalance causes sideways displacement of the stereo image. Balance is usually less than 2 dB.

Pickup arms

The function of carrying the cartridge across the disc is not as simple as it looks. Lateral tracking error must be minimised by optimum offset angle and overhang. Other ways have been used to avoid tracking error.

Tangential/parallel tracking arms   One method is to mount the head shell on several pivoted rods so that it swivels as it tracks across the disc thus always maintaining a tangent to it. However, the numerous pivots each contribute friction, so the arm needs a low compliance cartridge to enable it to track without pulling the stylus off-centre. Also each rod has its own resonant frequency which can fall within the audio range. Another type has a short arm driven along a carriage across the disc by a small motor. As the pitch of the grooves vary, some sensing device must be included to control the drive which adds to the complication and cost.

The distortion produced by a well designed conventional arm can be less than 1%, whereas that generated by vertical tracking and tracing error can be 7% or more. So lateral tracking error is the least problem, and its avoidance by tangential arms is outweighed by their disadvantages.

Arm resonance   The arm resonant frequency depends on its mass and its support springiness, which is the cartridge compliance. The formula is:

$$f = \frac{10^3}{2\pi\sqrt{MC}}$$

where M is the combined mass of the arm and cartridge in grams and C is the compliance in 10^{-3} N/m or 10^{-6} cm/dyne.

Output drops rapidly below resonance, so it should be pitched high enough to reduce lower frequency rumble, disc warps and ripples which can otherwise produce large amplitude subsonic signals that could overload the amplifier and even damage the speakers. Rumble and flutter can produce intermodulation sidebands at audio frequencies. Their amplitude depends on the resonant frequency, decreasing as it rises.

So a high resonant frequency is desirable, yet it must also be below the lowest recorded audio frequency of 20 Hz. A frequency only just below is therefore the optimum; 10 Hz was once recommended, but 16 Hz is now the preferred figure. A quick check is that *the product of the mass and compliance should equal 100 for a resonant frequency of 16 Hz*. For a frequency of 10 Hz the product is 250 which is the upper limit. High compliance is clearly undesirable unless mated with an arm of low mass.

In addition to the mass/compliance resonance, there is a *self resonance* that falls usually within 100–250 Hz. The main effect of this is a considerable increase in crosstalk between channels. It can be minimised by viscous fluid arm-damping and flexible decoupling of counterweights.

Tracking weight adjustment   A counter weight is slid along the rear arm extension, and clamped when the arm is in balance. Then a spring tension is adjusted to add the required tracking weight. The tension control may be calibrated in grams, or an external balance may be required.

Side thrust   Rotary friction pulls the stylus toward the centre of the disc even when it is blank. In the groove this means extra pressure on the inner wall. To offset this, *bias compensation* in the form of a small weight which is lifted on a string or pivoted lever by the inward arm travel, or by attraction from a weak magnet, is provided.

Construction   Lightweight construction is necessary to reduce mass, yet rigidity is also essential. Tubular aluminium, magnesium, and wood have been used, and the shells skeletonised. Low friction pivots require forces of no more than 20–50 mg maximum applied at the stylus, to move the arm horizontally or vertically. Unipivots are often used to achieve this.

Energy shunts   Vibration from the stylus spreads from the cartridge through the shell and the arm, but at each interface there is a reflection which eventually arrives back at the stylus in similar manner to sound reverberation. A system of energy shunts to bridge the intersections has been devised to counteract this effect. These are fine brass wires that are bridged across physical interfaces such as the head shell/arm to conduct vibrational energy across the interface and prevent it being reflected back to the stylus. They are produced by one small firm and so are not commonly used or well known.

Arm wiring   Arms are wired to a standard colour code which is:

LH live	LH screen	RH live	RH screen
white	blue	red	green

Load/lead capacitance   The inductance of the cartridge combined with the capacitance of the screened lead forms a resonant circuit that can produce a peak in the upper part of the frequency response. It can be flattened by critical resistive damping, or pushed above the audio range with low values of inductance and capacitance. As sufficiently low values are hard to get in practice, damping is applied using a load resistor.

Its value is effective for only one combination of inductance and capacitance, so as the inductance is fixed, and the load is standardised at 47 kΩ with few exceptions, the specified capacitance must be observed.

While some low-noise and data-screened cables have capacitances around 100 pF/m, most usually used for pickup connection go up to 350 pF/m. As some cartridge makers specify 100 pF, care must be taken in selecting and measuring the screened cable.

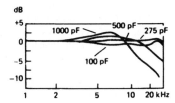

Figure 54. Response of a 700 mH, 47kΩ load, 275 pF optimum lead capacitance cartridge at various other capacitances.

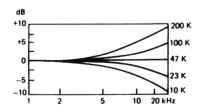

Figure 55. Same cartridge at 275 pF but with various other load resistances.

Turntables

Motors, synchronous/asynchronous In the *asynchronous* or *induction* motor, current is induced into the rotor windings. Speed is voltage and load-dependent which permits eddy-current braking control but has poor speed stability. The *synchronous* motor has a permanent magnet rotor that keeps step with a rotating field generated by the stator, by copper-ring shading or phase-splitting alternate poles. Speed depends on pole number and mains frequency. A 4-pole motor runs at 1500 r.p.m. at 50 Hz; and so has a vibration/rotation frequency of 25 Hz with harmonics at 50 Hz and 75 Hz.

Multi-pole motors Up to 24-pole motors have been produced, giving speeds down to 250 r.p.m. Rotational frequency is 4 Hz with harmonics at 8 Hz and 12 Hz, which is below the audio range. They also have a smoother torque.

Servo-d.c. motors A magnetic generator on the turntable produces a voltage which is compared with a reference and any discrepancy is electronically adjusted. Speed is thus constantly monitored and corrected.

Servo-a.c. motors A low voltage a.c. motor is run from an accurate oscillator. A toothed ring on the turntable generates a frequency that is compared to that of the oscillator which is varied to compensate for any speed difference.

Direct drive The turntable is part of the motor, having a magnetic ring clamped to its underside which follows the rotating field of the stator coils mounted beneath it.

Load variation/hunting Servo control requires low-mass motor rotors and turntables that can respond rapidly to speed correction. Even with these there is a lag in the response, with consequent overshoot, to both the start and finish of a correcting signal. This is termed *hunting*. Sudden speed variations can be caused with lightweight turntables by contrasting heavily and lightly modulated recorded passages. This can sometimes be heard as an increase in pitch in the dying reverberation after a loud musical chord. So with electronic control systems, turntable mass must be a compromise between rapid response and adequate momentum to resist load-variation speed changes. Synchronous motors with heavy turntables do not suffer from these problems.

Speed variations with load can be caused by the record slipping or lateral flexing of the mat. One system used a suction pump to create a vacuum between disc and turntable. A simpler alternative is a record clamp.

Drive systems A rubber *intermediate* or *idler* wheel was originally used to drive the inside of the turntable rim from a three-tiered motor capstan. Changing the drive from one tier to another changed speeds from 78 through 45 to 33 r.p.m. Mechanical noise from the motor, idler bearings and its two contact points were transmitted through the turntable to the stylus. (The common belief that the size of the idler wheel influences the turntable speed is untrue, the drive ratio from capstan to turntable is the same whatever its size.)

The use of a *belt drive* from motor capstan to a sub-turntable flange eliminates idler noise and decouples any vibration from the motor. The larger area of contact gives a more positive drive, especially when multipole slow-revving motors permit the use of a large diameter capstan.

Turntable parameters

Rumble Very low-frequency noise generated by the turntable bearing and the intermediate wheel if used. It is not usually noticed with equipment having a poor bass response. The quoted rumble figure is meaningless without a given reference level, however the usual reference generally understood is of a 1 kHz or 315 Hz signal recorded at 10 cm/s velocity.

The rumble measurement is weighted to include only the most annoying frequencies. Two weighting curves are in use: the DIN B which centres on 315 Hz, slopes either side at 12 dB/octave, and the more stringent DIN A which is flat from 12 to 315 Hz, sloping either side. Values lower than –50 dB for the A curve and –65 dB for the B can be considered good.

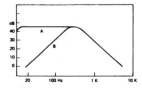

Figure 56. DIN A and B rumble curves.

Wow and flutter Wow denotes a slow, cyclic variation in pitch, while flutter is faster and irregular. Flutter can be encountered with servo-controlled direct drive motors and also due to the record/mat interface, but wow is the largest constituent of the specified figure. The peak value may be quoted, but it is usually r.m.s. because at 0.707 of peak it is lower.

The lowest detectable is 0.2%, but 0.05% is a typical specification.

The disc itself may contribute wow if warped. Also the centre hole may be up to 5 mil off-centre, and its diameter has a tolerance of 35 mil.

Turntable mat   1. Vibrations in the disc from the stylus are partly absorbed and partly reflected back at the mat interface. 2. Its flexibility can cause snatch and drag on heavily modulated passages. 3. Air trapped by the ribbing can form tiny helmholtz resonators. 4. It partly isolates the disc from turntable rumble and resonances. The extent of these effects varies with different materials, thicknesses and surface patterns, hence different mats can produce different results. The two optimum conditions are: total absorption so eliminating 1 and 4; or no mat at all thus disposing of 1, 2, and 3.

Acoustic feedback   The turntable can react to sound pressure waves from the loudspeakers which it conducts to the stylus. Resulting feedback can colour the reproduction or even produce a hoot. Feedback can also be conducted to the player through the floor and supports. Acoustic isolation of the player and the speakers is the remedy.

Stroboscope formula

$$N = \frac{f}{r}$$

where N is the number of markings; f is the flicker rate (100 for 50 Hz); r is the revolutions per second.

Compact discs

The compact disc offers many advantages over the LP record. Noise and distortion produced in recording and playback are eliminated. Dynamic range and channel separation is greater, and the disc is less vulnerable to damage because the information is not on the surface but is sealed in. Consequently, it does not deteriorate with repeated playings. It can store extra data in the form of programme details, expired time, and even two extra channels to give quadrasonic reproduction. Its 12 cm single side plays for 1 hour, which is longer than both sides of an LP.

Principles

The system uses digital techniques which were first patented by a British engineer Alec Reeves in the 1930s. Though impractical with components then available, they were revived in the late 1960s when the BBC developed digital links between radio studios and transmitters.

Analogue signals Signals recorded on conventional disc or tape vary in sympathy with the original sound pressure wave. With discs, the groove walls move in similar manner to the microphone diaphragm and with tape, the recorded magnetic zones vary in intensity. They are analogous to the pattern of the originals, hence the designation.

Any deformation of the signal or recording medium at any point during recording or playback produces distortion. Overloading can occur if the signal is too great, while it can be lost in noise if it drops too low. Dynamic range is thus limited.

Digital signals If a number of instantaneous measurements of an analogue waveform are made, the result is simply a succession of numbers each specifying the signal voltage after a given interval from the last. If those numbers are then used to produce a series of voltages at the same intervals as the measurements, the original waveform is reconstructed.

It matters not how the numbers are recorded, or whether there is noise or distortion in the recording; providing the numbers can be recognised by the reconstructing device the waveform will be identical. The only factors limiting the fidelity of the result are the adequacy of the number of samples measured, the accuracy of the measurements and the accuracy of the reconstruction. As the first two are standard parameters of the system, recording quality is uniform. The third is governed by the design of the player, so for a given design, the results are predictable and not subject to the vagaries of use.

Sampling frequency   The *Nyquist theory* requires a sampling rate of twice the highest frequency to be handled, which is thus represented by just two samples. This may seem insufficient. However, the highest frequency is bound to be a sine wave, otherwise harmonics consisting of still higher frequencies would be present. A sine wave can be represented by the values of each peak, so two are sufficient; but this assumes that the samples are taken when the peaks occur, which they may not. Having the sampling frequency a little higher, makes it sweep along the wave during successive cycles, sampling in different positions.

Audio frequencies higher than half the sampling rate produce errors known as *aliasing* so these must be eliminated with a steep top-cut filter. A slightly higher sampling frequency also allows for the upper roll-off of the filter. The sampling frequency used for compact discs is 44.1 kHz and the top audio frequency is 20 kHz.

Quantisation   As the samples are represented by numbers, these must be an accurate measure of the wave amplitude. This depends on the largest number that the system will accommodate, which in turn governs how many sampling levels can be used. If the maximum number was 100 this would represent the largest amplitude and there would be just 100 levels to quantify all those below it. A signal of –34 dB would therefore measure 2, and it would need an increase to –30 dB to reach 3. Thus intermediate values would be assigned the lower number and so be incorrect.

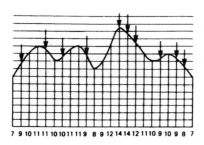

Figure 57. Quantisation. Measuring samples of an analogue wave in set levels. Arrows indicate intermediate values which produce distortion.

The waveform deformation caused by sample values falling between the measuring levels is termed *quantisation distortion*. As an error from 2.5 to 2 is greater proportionally than from 99.5 to 99 (20% and 0.005%), it follows that quantisation distortion is greater at lower signal amplitudes. The effect is that diminishing musical sounds decay in a series of steps instead of smoothly which gives them a 'crumbling' sound.

One method of reducing low-signal distortion is to arrange for the measuring levels to be closer together at lower amplitudes and wider spaced at higher ones. This is termed *non-linear quantisation*. It enables fewer levels to be used for a specified amount of distortion.

Quantisation economies   Quantisation levels multiplied by the sampling rate gives the amount of data a digital system must be capable of handling each second. Limited capacity systems may require economies. To reduce the sampling rate means curtailing the treble response. Reducing quantisation either increases distortion for the same dynamic range, or if the range is reduced it degrades the signal/noise ratio. Halving the quantisation increases noise by 6 dB.

One method of reducing quantisation developed by the BBC checks every 30 samples. There is very little waveform change over this number so a rough average is taken and represented by just two levels. The remaining levels are then used to give a fine adjustment to the individual samples.

Another system uses *delta modulation*. With this the waveform is sampled at a much higher rate, around 250 kHz, at which the change between samples is minute. Instead of recording the actual value, the system merely notes whether there is an increase or decrease from the previous sample. This can be recorded by using only one of two figures. Any under- or over-shoot is automatically corrected by the next sample. Resulting fluctuations are smoothed out by a top-cut filter at playback.

Compact disc quantisation   The compact disc uses 65,535 levels of measurement plus 0 giving 65,536 total, which is sufficient to enable linear quantisation to be used with low orders of distortion.

Binary code   If the numbers were recorded in decimal notation, each integer would need ten states to represent the digits from 0–9. This would require a varying signal that could suffer distortion and error just as an analogue signal does. There would thus be little advantage.

The coding least susceptible to error and most easily handled is one having only two states represented by a maximum and zero signal. This is achieved with the binary code by which numbers are expressed by the digits 1 and 0. The signal thus consists of a string of pulses and gaps and so is described as *pulse-code modulation*.

In the binary code each 1 starting from the right indicates a power of 2 commencing with 2^0 which is 1, followed by 2^1 which is 2; the next is $2^2 = 4$, $2^3 = 8$, and so on. The appearance of a 0 in any position indicates the absence of that power of 2 from the complete number.

Conversion, binary-decimal A binary number can be converted to decimal by listing its powers of 2 starting from the right and adding them. Some examples are:

111	$1 \times 2^0 =$	1	11101001101	$1 \times 2^0 =$	1
	$1 \times 2^1 =$	2		$0 \times 2^1 =$	0
	$1 \times 2^2 =$	4		$1 \times 2^2 =$	4
		7		$1 \times 2^3 =$	8
				$0 \times 2^4 =$	0
101100	$0 \times 2^0 =$	0		$0 \times 2^5 =$	0
	$0 \times 2^1 =$	0		$1 \times 2^6 =$	64
	$1 \times 2^2 =$	4		$0 \times 2^7 =$	0
	$1 \times 2^3 =$	8		$1 \times 2^8 =$	256
	$0 \times 2^4 =$	0		$1 \times 2^9 =$	512
	$1 \times 2^5 =$	32		$1 \times 2^{10} =$	1024
		44			1869

As a shortcut, count the number of digits to the right of each 1 reckoning each number as a power of 2, then add their values.

Conversion, decimal-binary To convert a decimal number to binary, divide it successively by 2, noting down the remainders (which must either be 0 or 1) writing from right to left. The final dividend is always 1 so as this cannot be divided by 2 it becomes the last remainder and final left-hand figure. As with decimal numbers, the least significant digit is on the right. Examples are:

23	$23 \div 2 =$	11>1	3324	$3324 \div 2 =$	1662>0
	$11 \div 2 =$	5>1		$1662 \div 2 =$	831>0
	$5 \div 2 =$	2>1		$831 \div 2 =$	415>1
	$2 \div 2 =$	1>0		$415 \div 2 =$	207>1
	$1 \div 2 =$	->1		$207 \div 2 =$	103>1
		10111		$103 \div 2 =$	51>1
				$51 \div 2 =$	25>1
74	$74 \div 2 =$	37>0		$25 \div 2 =$	12>1
	$37 \div 2 =$	18>1		$12 \div 2 =$	6>0
	$18 \div 2 =$	9>0		$6 \div 2 =$	3>0
	$9 \div 2 =$	4>1		$3 \div 2 =$	1>1
	$4 \div 2 =$	2>0		$1 \div 2 =$	->1
	$2 \div 2 =$	1>0			110011111100
	$1 \div 2 =$	->1			
		1001010			

Decimal/binary conversion table

Decimal	Binary	Decimal	Binary
1	1	60	111100
2	10	64	1000000
3	11	70	1000110
4	100	80	1010000
5	101	90	1011010
6	110	100	1100100
7	111	110	1101110
8	1000	120	1111000
9	1001	128	10000000
10	1010	130	10000010
11	1011	140	10001100
12	1100	150	10010110
13	1101	160	10100000
14	1110	170	10101010
15	1111	180	10110100
16	10000	190	10111110
17	10001	200	11001000
18	10010	256	100000000
19	10011	300	100101100
20	10100	400	110010000
21	10101	500	111110100
22	10110	512	1000000000
23	10111	600	1001011000
24	11000	700	1010111100
25	11001	800	1100100000
26	11010	900	1110000100
27	11011	1000	1111101000
28	11100	1024	10000000000
29	11101	1200	10010110000
30	11110	1400	10101111000
31	11111	1600	11001000000
32	100000	1800	11100001000
33	100001	2000	11111010000
34	100010	2048	100000000000
35	100011	2500	100111000100
36	100100	3000	101110111000
37	100101	3500	110110101100
38	100110	4000	111110100000
39	100111	4096	1000000000000
40	101000	4500	1000110010100
41	101001	5000	1001110001000
42	101010	6000	1011101110000
43	101011	7000	1101101011000
44	101100	8000	1111101000000
45	101101	8192	10000000000000
46	101110	9000	10001100101000
47	101111	10000	10011100010000
48	110000	15000	11101010011000
49	110001	25000	110000110101000
50	110010	60000	1110101001100000

Notes: (1) a single 1 followed by 0s is always a power of 2, the power being given by the number of 0s; (2) to double a number add a 0 to the right; to halve one remove a 0; (3) as with decimal numbers, all numbers ending in 0 are even.

Maximum values   The maximum value for a given number of digits is when all of them are 1s. The maxima for up to 16 digits are:

digits	decimal max	digits	decimal max
1	1	9	511
2	3	10	1023
3	7	11	2047
4	15	12	4095
5	31	13	8191
6	63	14	16383
7	127	15	32767
8	255	16	65535

Gray code   This is a variant of the binary code. Its feature is that only one digit changes each time instead of one or more with binary. Thus false data is less likely from voltage spikes during conversion.

Decimal	Binary	Gray	Decimal	Binary	Gray
1	0001	0001	9	1001	1101
2	0010	0011	10	1010	1111
3	0011	0010	11	1011	1110
4	0100	0110	12	1100	1010
5	0101	0111	13	1101	1011
6	0110	0101	14	1110	1001
7	0111	0100	15	1111	1000
8	1000	1100			

Compact disc format

Physical details   The disc consists of clear polycarbonate plastic. Data is impressed into one side in the form of a spiral train of microscopic pits which is silvered and covered with a protective coating. Information is read by a laser beam from the other side through the material. There is thus no physical contact with the recorded data during playing or handling. The recorded track starts inside and ends on the outside of the disc. Track velocity is constant at 1.2 m/s, so rotation speed is variable, slowing as the recording proceeds.

Disc data

Diameter	120 mm (4$\frac{1}{2}$ inches)
Thickness	1.2 mm
Diameter centre hole	15 mm
Programme start radius	25 mm
Programme finish radius (max)	116 mm
Rotation	Anti-clockwise to laser
Rotation speed	500–200 r.p.m. (approximately)
Maximum recording time	60 minutes
Channel number maximum	4
Track pitch	1 6 µm
Pit width	0.6 µm
Pit or land length per digit	0.3 µm
Pit depth	0.12 µm

Pre-emphasis The audio signal may have a pre-emphasis of 15 and 50 μs, if it has, the information is included in the recorded control and display code which switches in a de-emphasis circuit.

Encoding The 65,535 signal levels are represented by a 16-digit binary number called a *word*, which is split into two 8-digit *symbols*, each digit being termed an *audio bit*. The LH and RH stereo channels are two consecutive words which constitute one *sample*. Six samples are grouped into a *single frame* which thus comprises 12 words of 24, 8-bit symbols.

Eight, 8-bit parity symbols are added to the frame for error correction, four after each group of twelve audio symbols. In addition, an extra 8-bit symbol is included for control and display information at the start of the frame, which with these now contains 33 symbols. The bits of the combined audio, parity and control symbols are called *data bits*.

Next, the symbols are converted from 8-bit to 14-bit units, this being known as *eight-to-fourteen modulation* (EFM). Three merging bits are added to bring the total to 17 bits per symbol. Finally 27 bits are added at the end of each frame for synchronising, making a total of 588 bits per frame. There are 7350 frames per second giving a channel bit frequency of 4.3218 Mbits/s. These are now termed *channel bits*.

Frame format

6 samples of 12 words of 16 bits =	24 x 8	audio bit symbols
parity	8 x 8	bit symbols
C and D	1 x 8	bit symbols
	33 x 8	data bits
EFM	33 x (14 + 3) =	561 channel bits
synchronising bits		27
Total channel bits		588

C and D Although there is only one control and display symbol for each frame, there are 7350 of them each second so information as to the title of the music, composer, timing and track number can be included as well as technical instruction to the player regarding de-emphasis.

Error correction The sums of sequences of numbers can always be made even by adding a 1 to the odd sums, and a 0 to the evens. Should any later add up to an odd number, an error is indicated. This is known as *even parity* and the added digit a *parity bit*. With binary coding where the digits are 0s and 1s an error involves the loss or addition of a 1.

Parity blocks The erroneous digit can be identified by forming the sequence into a block of digits, then the sums of both the rows and the columns can be made even by adding a parity bit to each row and column. An error makes one row and also one column odd, so it can be traced to the digit common to both at their intersection.

Distance A single error changes the information bit, and row and column parity bits, which is 3 in all. The block is thus said to have a *distance* of 3 (the incorrect block is distant by 3 digits from the correct one}. So the minimum distance is $2t + 1$ where t is the number of errors.

Blocks of symbols Blocks are formed from groups of eight digits called symbols, instead of from individual digits. The object is to achieve the maximum correction with the minimum parity symbols. To spread a string of errors over a large area and so make correction easier, alternate symbols from each frame are interleaved with those of the next frame.

Erasure positions The number of correctable errors by the decoder is $(d_m - 1) \div 2$ in which d_m is the minimum distance. But if the positions of the faulty symbols in the block are identified and labelled in advance, the decoder can erase them and insert calculated correct values. The errors that can be so corrected is $d_m - 1$ or twice as many as before.

Correction After removal of the C and D symbol the 32 symbols are fed in parallel to 32 inputs on the first *cross-interleaved Reed-Solomon code* decoder via delay lines in alternate ones that restore the interleaved symbols to their respective frames. The decoder corrects one faulty symbol but if there are any more they are not corrected, but all are given an erasure flag and passed on. The 28 audio symbols are then sent to the second decoder via delay lines of varying length to spread errors over several frames. Up to four faulty symbols can be corrected here; over that, the 28 are passed out uncorrected but still with their flags.

Interpolation Next, flagged symbols are erased, and their value estimated from adjacent unflagged ones. The odd and even RH and LH channels within each frame are interleaved, enabling up to seven consecutive samples to be interpolated. As the second decoder delay lines space the successive samples eight frames apart, this is multiplied to 56 frames, Further 2-frame delay lines enable a complete frame to be substituted if it evades previous correction. This reduces the 56-frame capability by 2 and the recording delay lines further reduce it by 5 plus. Reliable interpolation is thus up to 48 frames. Distortion is increased and bandwidth reduced to 10 kHz over the interpolated section. Maximum correctable burst length is 4000 data bits (2.5 mm on disc), maximum interpolation 12,300 data bits.

Muting If a stream of uncorrectable errors occurs the gain is dropped from 32 samples before to 32 samples after it. The gain variation follows a cosine curve to avoid harmonic generation produced by a sudden change.

Information density The number of channel bits that can be recorded depends on the diameter of the light spot, as this determines the minimum size of bit it can resolve. The spot does not have a sharp boundary, so the rated diameter is that at which the intensity is at half value. It is:

$$d = \frac{0.6\lambda}{NA}$$

in which d is the spot diameter; λ is the laser wavelength; and NA the numerical aperture of the objective lens. For a wavelength of 0.8 µm and NA of 0.45, the diameter is 1 µm.

Tolerances and aperture Although a smaller spot size could be achieved with a larger NA, this decreases manufacturing tolerances below practical limits. Some of these and their relationship to NA are:

Item	Tolerance	Proportion to NA
Lens tilt	±0.2°	NA^{-3}
Tracking	±0.1 µm	
Focusing	±0.5 µm	NA^{-2}
Disc thickness	±0.1 mm	NA^{-4}
Disc flatness	±0.6°	
Pit edge position	±0.05 µm	
Pit depth	±0.01 µm	

Eight-to-fourteen modulation The digital 1s and 0s are not represented by pits and spaces (termed *land)* as is sometimes stated. A 1 is indicated by either a leading or lagging *pit edge*. Pits and land could thus be interchanged on the disc without affecting the coding

As the light spot cannot read two edges at the same time the pit length cannot be shorter than the spot diameter. So, when representing a number having alternate different bits (01010...) the bits must be equal to or longer than the spot diameter. With a number having two or more adjacent 0s (10010001...) the bits can be shorter, in this case the pit length (or land between pits) denotes the number of 0s between the 1s.

If all numbers are like that more bits can be accommodated in a given space. This is done by converting the original binary number in each symbol into a different number which has two or more adjacent 0s. A 'dictionary' of logic gates is used to accomplish this in recording, and a similar one in the player translates back to the original binary numbers. The 8-bit symbols become 14-bit in the process, but even with these extra bits the system permits 25% more information to be recorded.

Minimum and maximum   To be effective there must be a minimum of two consecutive 0s, but there is also a maximum. There is no signal change during a long run of 0s, so the internal clock which is controlled by synchronising bits could get out of sync. The longer the run, the lower the frequency of the light beam modulation. As the servo systems operate at low frequencies (below 20 kHz, a long run could confuse them. The maximum number of consecutive 0s is therefore set at 10.

Merging bits   A symbol ending in a run of 0s followed by one starting with a run could exceed the maximum. Also a symbol ending in 1 or 10 would not produce the minimum two 0s if the next symbol started with 01 or 1. Three merging bits are added between symbols to avoid this, and are discarded after decoding. If more than one combination satisfies the conditions, the one that brings the lowest *digital sum value* is chosen. That is the ratio of pit and land length from the start of the disc. Minimising it reduces noise in the servo frequency band.

The CD player

Spot size   Light source for disc scanning is a laser because it generates a coherent beam of a single light wavelength that can be focused down to the 1 µm spot size required. Any other source contains multiple wavelengths that separate when passed through a lens to produce *chromatic aberration*, which is the appearance of various coloured outlines around the focused image. Achromatic lenses that avoid this are made of two elements of different types of glass, one cancels the dispersion of the other, but to make such a lens with the ability to focus to such a fine spot would be very costly.

Interference modulation   Light reflected from the pits, or bumps as seen from the light beam side, is a quarter wavelength different from that reflected from the adjacent land. A phase difference thus occurs which produces partial cancellation and darkening. Much of the remaining light is diffracted by the bumps and scattered because their width is smaller than the light wavelength. These effects which produce the dark modulations of the beam could only be achieved with light of a single wavelength, that is from a laser.

Laser   Lasers used for CD players are solid state aluminium gallium arsenide units having a wavelength of 0.8 µm and operate at low voltage. Life expectancy has been improved to 25,000 hours.

Basic functions   The laser beam reflected from the disc must be modulated by the pits, but having no physical contact it must keep itself centred on the track, and must maintain focus over any variations of the disc surface. There are thus three functions to perform. Two basic methods exist to accomplish these each having variants. These are the three-beam and single-beam systems.

Three-beam system

Optical path   From the laser, the beam passes through a *collimator lens* to ensure that it is parallel, then through an *optical grating* which splits the beam into three. The main one contains 50% of light, while the others on either side of it contain 25% each.

From there, the beams pass through a *Wollaston prism* which performs a vital function on the return journey, and then through a quarter-wavelength plate. This device produces a $45°$ rotation in the plane of polarisation and is also necessary for the return pass.

Next, the beams are directed up through the focusing lens to the underside of the disc. On reflection, they travel back through the lens and again encounter the quarter wavelength plate. There the polarisation is given a further $45°$ rotation, which means they are now at $90°$ polarisation to the forward beams.

Next comes the Wollaston prism, which is a three-element quartz device that produces refraction angles that are dependent on the plane of polarisation. Thus the beams are diverted from the forward path at this point and pass through an *astigmatic lens* that is required for the focusing function. From there they are directed to the photodiodes where the light and dark modulations are converted into an electrical signal.

Tracking with three beams   Beam-tracking error must be within 0.1 μm, so the tracking motor which moves the pickup assembly across the disc must be very accurately controlled by the track itself. The two auxiliary beams are positioned 20 μm before and 20 μm behind the main one, one offset to the left and the other to the right.

These beams thus read along the edges of the track and when correctly centred they return equal illumination to their respective photodiodes. If the assembly goes off-centre, one tracking beam will read more of the modulations while the other reads more of the flat surface. Illumination is no longer equal and there are unequal outputs from the photodiodes. Outputs are compared to produce an error signal for continually correcting the motor in the appropriate direction.

Should a blemish on the disc surface temporarily obscure either of the tracking beams, the correcting signal hence the motor stops and the pickup remains stationary under the one track until it picks up the signal further on.

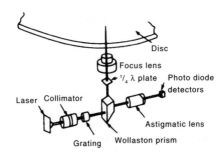

Figure 58. Optical path for three-beam system.

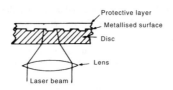

Figure 59. Refraction of beam through disc material lengthens field of focus while wide beam area at surface minimises effect of blemishes.

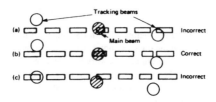

Figure 60. Disposition of tracking beams.

Focusing The depth of field with a lens of 0.45 aperture is 4 µm which makes focusing critical but it is helped by passing through the disc itself. It has a refraction index of 1.5 which is the ratio of the speed of light through the material compared to that through a vacuum. This produces refraction, bending the rays and lengthening the field of focus. It is still very small but this has one advantage in that the surface of the disc, being 1200 µm from the point of focus, is a long way out of focus, so any surface blemishes or dirt has little effect on the focused spots unless they are very large.

Focusing error must be within 0.5 µm, but displacements of up to 500 µm can be caused by disc warp, so focus must be continuously controllable. The lens system consisting of up to four elements, has a concentric coil surrounding it and is free to move vertically between the pole pieces of a magnet, The arrangement is similar to the cone of a loudspeaker.

The main signal beam falls at the centre of four photodiodes arranged in square formation. When in focus, the spot is circular thus illuminating equal parts of all four receptors. If it goes out of focus, the spot becomes elliptical due to the presence of the astigmatic lens in the optical path. When the focus is too close, the ellipse lies across the second and third quadrant, but when it is too far, it is across the first and fourth.

The four photodiodes are connected so that output of the diagonals add and each pair is compared with the other. An elliptical spot produces a higher output from one pair, so the comparator produces an error signal to apply to the focus lens coil. Polarity of the error signal, hence the direction of the lens movement, depends on which diagonal output is greater. All four diodes are summed to produce the main p.c.m. signal.

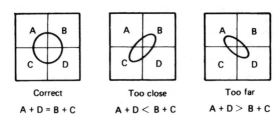

Correct	Too close	Too far
$A + D = B + C$	$A + D < B + C$	$A + D > B + C$

Figure 61. Out of focus elliptical beam illuminates diagonal pairs of photodiodes unequally. Direction indicates state of focus.

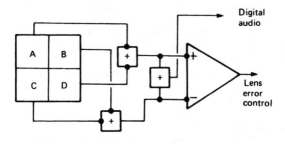

Figure 62. Diagonals are added and compared with opposite diagonals. All four are summed for the main p.c.m. signal.

Single-beam systems

If the two tracking beams are dispensed with, the optical system can be simplified. The optical grating, Wollaston prism and quarter wavelength plate can be eliminated. The complete optical system can be contained in a simple, easily replaceable plastic-cased unit which can be swung across the disc on a pivot like a pick-up arm, whereas a linear carriage is needed for three beams to maintain their tangential alignment. Furthermore, the whole energy of the laser is applied to the signal beam instead of 50% as with three beams. There are two methods of focusing and tracking.

Optical path   The laser beam is reflected by an angled half-silvered mirror through the focus lens. After reflection from the disc it re-enters the lens, passing through the mirror and final lens to the photo diodes.

Quadrant diode system   With this the focusing is the same as with the three-beam system, but the four quadrants of the detector are also used to detect deviation of the spot from the centre of the track.

If the beam is off-course, one half of the spot reads the pits while the other reads the land, but as the quadrants are connected diagonally for focusing correction, the output from each pair is the same. Both have one quadrant reading track and one reading land. The difference between each pair is the timing. The top quadrant of one pair reaches a pit before the bottom quadrant of the opposite pair.

So the output from one pair of quadrants arrives at the comparator just before the other, which results in a pulse each time the spot reaches or leaves a pit. The polarity of the pulse depends on which pair is affected first and therefore indicates to which side of the track the spot has deviated.

Thus an error signal is produced which is used to control the tracking motor. The pulses could interfere with the focus system, so a filter must be included to eliminate them from the focus error circuit. When the spot is on line, the pit reaches the top quadrants of both pairs simultaneously, and no pulse is generated.

In-line diode tracking   The difference with this system is that the beam is split into two by the final prisms in the optical path. Two prisms are combined, the exit one having a wedge shape which divides the beam. Their joint surface is half-

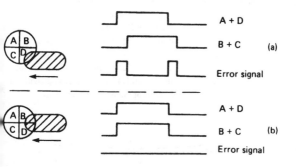

Figure 63. (a) as a pit reaches an off-line spot it encounters quadrant D before C, so that pair AD is affected before pair BC. A pulse is thereby produced from the comparator which serves as a tracking error signal; (b) When the spot is on-line A and C are affected simultaneously.

silvered by an evaporated film that serves to reflect the incoming laser beam up through the focus lens. Although split into two, the system is single beam because only one beam is applied to the disc.

Both halves of the beam fall on four photo diodes which are arranged in a row, One half fills equally between D1 and D2, and the other between D3 and D4. If the spot on the disc is off-centre, the half beam reflected from the spot area reading the land is brighter than that obtained from the part reading the pits. So there is a difference in output level between the diodes D1 + D2, and D3 + D4, and this is used as an error correction signal.

However, dirt on the lens or other defects can produce permanent brightness inequality between the two beams resulting in tracking error. To avoid this a second tracking error signal is generated by applying a 600 Hz current to the tracking coil. This makes the arm oscillate from side-to-side, displacing the beam by ±0.05 µm. Thus the two half-beams are modulated by a 600 Hz signal whose amplitude increases in one pair of diodes but decreases in the other if the spot tracks to one side. Summing the signal from both pairs produces either a positive or negative control signal depending on which side of the track the spot has strayed.

Focus When the spot is sharply focused on the disc, two sharp images appear on the photodiodes, one in between D1 and D2, and the other between D3 and D4.

If the spot goes out of focus, the images also become diffuse and move closer together or further apart. Thus the inner pair of diodes D2 and D3, have a different illumination level than the outer pair D1 and D4. A comparison provides a difference error signal that is applied to the focus coil.

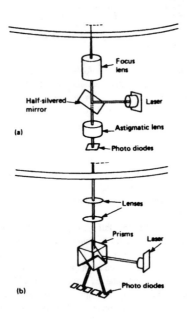

Figure 64. Single beam optical path: (a) quadrant diode system; (b) in-line diode system.

Decoding the signal

Demodulator and buffer The synchronising bits identify the start of each frame from the data bit stream from the photodiodes, which is then demodulated, converting the 14-bit symbols back to 8. Next the frames are fed into a buffer memory and extracted from the other end. The input rate may vary slightly due to disc speed irregularities, but the output rate is regulated by a quartz-controlled clock so no wow can appear in the reproduced signal. The buffer is kept nominally 50% full to permit regulation in both directions, but the amount in the buffer at any instant is an indication of whether the motor is slow or fast, so it is used as a correction signal to control the motor.

Correction and separation From the buffer, the C and D symbol is directed to the control and display circuits while the audio and parity symbols are fed to the error correction circuit. There the interleaved symbols are also restored to their correct frames. Then the audio symbols go to the interpolation and muting circuits to deal with errors too large for correction. The alternate RH and LH channel words are separated and applied via filters to the two digital-to-analogue converters.

Removing the steps After conversion, the wave is in the form of steps at the sampling frequency. To restore the original waveform, these need to be smoothed out. Each has a flat top, which consists of the sampling frequency and its harmonics. A top-cut (low pass) filter can remove them but must not affect the highest audio frequency; so it must fall within 20 kHz–44.1 kHz. The cut must be of at least 50 dB, and to achieve this without affecting frequencies only an octave below it needs a complex multi-order filter that can produce oscillatory 'ringing' at lower audible frequencies. This filter is one of the big problems; one solution is oversampling the signal before digital-to-analogue conversion.

Figure 65. Analogue wave and step produced by digital-to-analogue converter.

Oversampling   With some players, each digital sample fed to
the converter is electronically repeated. This cannot affect the
original sampling rate or the audio frequency response, but it
doubles the converter sampling frequency. In some cases there
is a second repeat producing four times the converter sampling
frequency. This is termed *two-times oversampling and four-
times oversampling* respectively. Sampling frequencies are thus
88.2 kHz and 176.4 kHz. As there is a much wider gap
between these and the highest audio frequency, simpler filters
can be used with gentler slopes which do not adversely affect
the audio spectrum.

If each original sample was merely repeated the original
steps would remain unchanged, so the repeats must be altered
to give intermediate values between adjacent original samples.

Transversal filter   One method of accomplishing this is by
means of a transversal filter comprising 24 series delay units
which each delay the signal by one sampling period. Part of the
output of each unit is passed along the line and the other part is
sampled four times per period. The samples are multiplied by a
coefficient and sent to an adding circuit with all the others.

The coefficient is a 12-bit word which is chosen so that
when it multiples the 16-bit sample, it produces a 28-bit word
and no extra bits are required when all the products are added.
The coefficient is thus constantly changing depending on the
previous 24 samples and so is able to calculate the three new
intermediate values between each received sample.

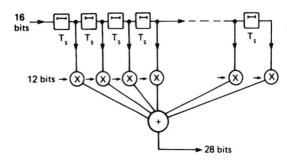

Figure 66. Transverse filter consisting of 24 delay elements
each holding one recorded word which is sampled four times,
each multiplied by a changing coefficient which depends on
the overall level of the 24 words.

14 or 16 bits   Early 16-bit decoders were quite inaccurate, often tending to ignore the least significant bit or two. Thus their resolution was equivalent to a 15-bit or 14-bit converter, but less accurate because the errors were not consistent. Some makers preferred to use accurate 14-bit converters. Furthermore as these were faster they allowed the practice of oversampling for which the then available 16-bit units were too slow.

Quantisation noise   The quantisation steps produce noise in the sampled frequency band which is proportional to the size of the steps. If the step size is doubled by reducing the bit number by one, the noise voltage is also double so increasing it by 6 dB. So a 14-bit system at –84 dB, has 12 dB more noise than a comparable 16-bit system at –96 dB.

With four times oversampling, the noise is spread over four times the frequency spectrum so that contained in the audio range is about a quarter of the power which is 6 dB less, giving a noise level of –90 dB.

Noise shaping   When rounding off to 14 bits there is a rounding off error which is almost the same for successive samples at low frequencies. Thus if the error is changed in sign and added to the next sample, the errors virtually cancel. At a sampling rate of 176.4 kHz, most of the audio spectrum is at relatively low frequency. So, considerable reduction of the error is achieved. This further reduces the noise by some 7 dB thereby giving a total noise level of –97 dB, which is slightly better than for a 16-bit system without oversampling.

Digital-to-analogue conversion   There are several methods of digital-to-analogue conversion. One uses a chain of resistors, each being twice the value of its predecessor, that are connected as a potential divider across the supply voltage. Different points of the chain are switched in turn to the output circuit by an array of electronic switches which are actuated by the bits of the digital signal. Thus the output consists of a series of voltage steps, each of a value determined by one of the binary words.

Another type uses a constant reference current that is divided by two, down through successive stages. These currents are periodically interchanged so that small differences are averaged out, a process termed *dynamic element matching*. The changes which are initiated by the crystal-controlled clock function down to the four least significant bits. These are divided passively without interchange. This type needs a *hold* circuit to keep the current constant at the last value until the next sample arrives, otherwise the output is a series of sharp pulses instead of steps. The low-pass filter then rounds off the steps.

Player specifications   Most players fall within these sets of figures.

Frequency response	Dynamic range	S/N ratio	Channel separation	THD	Wow	Output
2–20 kHz	>90 dB	>90 dB	>90 dB	0.0015%	_	2 V
20–20 kHz	>90 dB	>90 dB	>90 dB	0.05%	_	2 V

Some noise is contributed by the power supply and the analogue output stage, so it is marginally greater than the noise calculated from the quantisation parameters.

Rumble/acoustic feedback   Mechanical noise is communicated from the motor and drive components via the turntable to the pickup stylus in the case of the analogue gramophone. As there is no physical contact between the disc and the pickup device there can be no rumble. Similarly, acoustic feedback from loudspeakers has no effect on the CD player.

Four-channel stereo   The compact disc is capable of containing four separate channels if required. These would be sampled consecutively in the same manner as the present two, but more space would be taken on the disc and so playing time would be reduced.

CD-ROM   The compact disc has a capacity of 600 megabytes of data memory, equivalent to 1800 floppy discs or 250,000 typed pages of A4. Access time is some $1\frac{1}{2}$ seconds. The standard for CD-ROM contains sectors that can be addressed with an absolute sequence number starting from the beginning of the track. This can be in the number of minutes, seconds and $\frac{1}{75}$th of a second (the period of one frame).

CD-I   CD Interactive combines text, data, still pictures and diagrams plus sound from top quality to speech. All can be used simultaneously on players that can also play conventional music CDs.

Ageing   Fears have been expressed as to whether the plastic will cloud with age and so prevent penetration by the laser beam. Makers state that polycarbonate does not suffer from this form of deterioration. Some cases of oxidisation of the aluminium coating have been reported due to inadequate sealing. Makers are improving the lacquer seal, and gold-plating instead of aluminium has been suggested.

Compact disc manufacture

Recording   To get full benefit from CD records, the master tape should be digitally recorded, because even studio analogue recorders are inferior to CD systems in respect of noise and dynamic range. Yet many CD records are made from analogue masters.

The master tape cannot be edited by cutting and splicing like analogue tape, but it can be copied for many generations with no deterioration. So editing consists of copying the desired parts on to another tape which then becomes the master.

Master disc   A disc of highly polished optically flat glass is coated with a layer of photo-resistant material. Scrupulous cleanliness must be observed at all stages as one speck of dust could obscure a large part of the recording. The blank disc is taken to the recording machine in a sealed container. There it is recorded with a modulated laser that etches the pits in the layer. Excess resist material is chemically cleaned off and the disc is silver plated before being nickel plated.

Fathers and mothers   The nickel is built up then separated from the master disc to form a *negative father* having bumps in place of pits. A number of positive *mothers* are made from this. Further electroplating of these mothers produces a series of negative *sons* that serve as the actual stampers.

Stamping   The disc can be stamped out or injection-moulded. With the latter, a pellet of polycarbonate plastic is inserted automatically into the press where it is pressed into the finished shape with the billions of modulations. This takes about 20 seconds.

Silvering   From the press, the disc is cooled and taken to a vacuum chamber with hundreds of others. Here they encounter a mist of ionised aluminium which bonds to the modulated surface of the discs, depositing a layer of some 0.04 µm thick. This process can take about half-an-hour, most of which is needed to pump the high vacuum required. A different approach is to silver each disc individually in a large number of small chambers. These take only about four minutes to evacuate.

Coating   The next process is to place a spot of lacquer on the silvered surface and spin the disc to distribute it evenly. Thus the surface is sealed and protected.

Centre hole   Up to now the disc has had a small pilot hole in the centre, Punching the final hole must be done with extreme precision; an eccentricity of 0.1 mm could cause the reading laser beam to wander over 60 adjacent tracks! The optically correct centre (the centre of the circular tracks, not necessarily of the disc perimeter) is found and the hole punched by laser beam. In one plant, if an accurate centre cannot be found after three attempts by the laser, the disc is scrapped.

Finishing A special offset printing press prints the programme details on the lacquered side of the disc, which is then carefully examined under polarised light and packed.

Rejects There are some 60 points during the manufacture at which losses can occur. Total rejection from all causes runs at around 15%. As polycarbonate cannot be re-cycled, it is ground up for building material.

Compact disc faults

Many of the faults that are reported with CD players are due to the disc itself. Some of these can be misleading and suggest that the fault lies with the player. Confusion can result when a disc plays perfectly on one player but produces a fault on another. Some machines seem to take a dislike to particular discs.

The effect should not be unexpected, as similar things happened with LP records, with groove jumping and distortion being experienced with a particular disc on one player, but not another.

Surface faults There is a popular misconception that CDs are indestructible, no doubt produced by over-optimistic maker's claims when CDs were introduced. Minor scratches are largely ignored due to the system's error compensation, but there is a limit, and many faults can be produced by more serious abrasions.

If the scratches render the disc unplayable, there are two possible remedies which may restore it. One is to polish out the scratches with a fine abrasive cleaner, the other is to paint the scratched area with a thin clear varnish using the minimum of varnish and avoiding leaving brush strokes. The latter is the easiest and may prove the more successful.

Dirt may not be immediately noticeable but can produce various faults. Cleaning can be done with a dry cloth or, if the dirt is stubborn, with a fluid formulated for the purpose. Cleaning strokes should be radial from centre to perimeter otherwise the dirt is deposited elsewhere.

Silvering faults If moisture gains access to the silvering, it can cause discoloration just as with an ordinary mirror. It is revealed by careful visual examination under a bench lamp. If it is bad enough to affect the playing, there is little that can be done.

Damage to the label side of the disc may expose and remove part of the silvering. The effect depends on the machine but it could range from loss of programme to ejecting the disc. An application of silver paint may help if the modulations are not damaged, otherwise the disc is useless.

Centre-hole faults Burs or runs of sealing lacquer inside the hole can make a tight fit on the spindle causing it to stick. Visual examination under magnification should reveal the trouble.

A worn centre hole results in eccentric running. Minor deviations can be dealt with by the tracking system but larger ones may not. A good test tool is a gauge consisting of a rod exactly 15 mm in diameter tapered or chamfered for ease of insertion. It should make a fit without forcing, and with no play.

Player specifications A disc may be slightly sub-standard in some respect but be within the range of the appropriate player compensating system to correct. However, play correcting abilities vary, so it may play well on one machine but not on another. If no other disc gives problems, it may be assumed that the disc is outside its tolerance, and the second machine has a better correction specification. Test discs such as the Philips SBC421 and SBC426 can check player specifications against the maker's figures quoted in the manual, and removes the uncertainty.

The Sony Mini Disc

Designed for the portable audio market, the Mini Disc features small size, ($2^1/_2$ inch diameter contained in a 72 x 68 x 5 mm caddy — compared to the $4^1/_2$ inch diameter of the normal CD), it also withstands signal loss due to physical shock while playing. In addition, the discs can be erased and re-recorded unlike the standard CD.

CD similarities   While much of the signal processing is similar to that of the CD, there is much that is new. Firstly, we will list the similarities which have been described in the previous pages. The analogue signal is converted to digital using 16-bit linear words and a sampling frequency of 44.1 kHz. This gives a frequency response up to 20 kHz, and the dynamic range is 105 dB. Eight-to-fourteen modulation is used along with the Reed-Solomon error correction system. The disc rotation speed varies to give a constant track-reading velocity of 1.2–1.4 m/s.

ATRAC   Although the disc size is smaller, and the sampling frequency the same, the playing time is actually greater than the CD, being a maximum of 74 minutes. This is achieved by Adaptive Transform Acoustic Coding (ATRAC) which is based on the same two psychoacoustic principles of human hearing as the PASC coding system used with the Philips digital compact cassette (see page 151).

The first of these is the threshold of hearing. The bottom contour in Figure 3 (see page 3) depicts the threshold of hearing. Anything below this is inaudible, so there is no need to record it. The second principle is the related one of masking. When a loud sound is heard, the threshold of hearing is raised at that point, so that another quieter sound of similar frequency which by itself is above the threshold hence is audible, now lies below it and so becomes inaudible. Such sounds also do not need to be recorded.

A 'snapshot' is taken of the audio signal by the encoder every 20 ms, each containing just under 1,000 samples. These are subject to a Fourier analysis to resolve their component sine wave fundamentals and harmonics. Any components that fall below a predetermined level are rejected and only those above it are encoded. Thus any redundant signals that fall below the hearing threshold whether basic or as a result of masking, do not take up space in the recorded data.

The Philips PASC system does something similar but divides the audio spectrum into 32 third-octave bands and if only signals below the predetermined level are found in any band, that entire band is suppressed. ATRAC is claimed to be five times more efficient than the standard CD in terms of data storage.

The Mini Disc   The recording and playback system is actually magneto-optical so is quite different from the CD which is purely optical. The disc has a magnetic layer of terbium ferrite cobalt, but instead of magnetic zones lying end-to-end as they do in magnetic recording tape, they are positioned vertically so take up much less space. The arrangement is like packing bar magnets upright instead of laying them on their sides. The change from a binary one to zero and vice-versa is achieved by a reversal of the magnetic polarity. This arrangement produces less self-erasure than with the lateral zones of magnetic tape, as well as a greater packing density.

Recording   Recording is by a magnetic head that tracks across the disc, and is assisted by a laser beam of fixed power. This heats the magnetic layer at the point of recording to above the Curie point of the material at which it becomes paramagnetic, that is, it loses any existing induced magnetism, but is highly susceptible to fresh magnetisation. Thus any previous recording is erased and the new recording implanted.

A very low recording flux of 80 oersteds is applied. This is sufficient to change the magnetic flux orientation in the susceptible heated area, but not elsewhere. Thus the laser beam effectively limits the effect of the magnetic field to a very small spot, thereby achieving a high packing density recording.

Replay   When a polarised beam of light enters a magnetic field its polarisation plane is rotated, either clockwise or anti-clockwise depending on the magnetic polarity. This is termed the Kerr effect. During replay, the laser beam is directed onto the disc and it tracks the train of magnetic zones. It is reflected back through a lens system and beam splitter which directs it onto a pair of photo diodes in a similar way to that used in the conventional CD player. As the magnetic poles reverse, the reflected beam polarisation changes, and one diode receives more light than the other. The electrical output of these is subtracted to give either a positive or negative result, which corresponds to either a zero or a one binary number.

Address   Address information is included after every 573 signal samples which is every 13 ms. This can be used to precisely locate passages on the disc, but is also necessary for the shock-proofing feature.

Shock-proofing   The data transfer from the disc is at the rate of 1.4 Mbits/s. However the decoder requires a data rate of only 0.3 Mbits/s so the data can be stored by a RAM buffer before reaching the decoder. The buffer is of 1 Mbyte and it stores 3 seconds of signal.

Should there be an interruption of the signal from the disc due to physical shock or other reason, the stored signal will continue to be supplied from the buffer to the decoder.

Meanwhile the tracking mechanism searches for the last received address mark on the disc, and when it is located, the data stream is again fed into the buffer. Thus there can be an interruption of up to 3 seconds with no audible effect on the output.

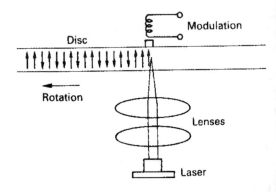

Figure 67. Magneto-optical system of the Sony Mini Disc.

Tape recording

Magnetism

There is often confusion over the principles of magnetism and the terms used to denote its various properties. This has not been helped by the changing of some of the units from c.g.s (centimetre, gram, second) to the clumsier MKS (metre, kilogram, second), resulting in both old and new terms being encountered. These need to be understood in order to grasp the terms used in connection with tape recording.

Magnetic moments   A magnetic field is produced when an electric current, which consists of electrons, flows in a circuit. The motion of orbiting electrons in the atom thereby generates a magnetic field, with further fields being produced by the electron spin. These are termed the *magnetic moments* and the moment of the whole atom is the vector sum of the individual ones.

   The sum may be zero in which case the material is termed *diamagnetic*, being non-magnetic. If there is a remaining moment, the material is said to be *paramagnetic* and can exhibit magnetic properties.

Magnetisation   As opposite magnetic poles attract each other, the molecules of a paramagnetic substance arrange themselves in circles so forming complete magnetic circuits. Thus the fields are all contained within the material and none appears outside.

   When such a material is placed in a magnetic field, the internal fields line up under its influence and produce an external field which reinforces the applied one. If when the applied field is removed, the molecules revert to their former state, the induced field disappears and the material is said to be magnetically *soft*. If some remain in line, these continue to exhibit an external field, and material is termed magnetically *hard*. The magnet so produced is called a *permanent magnet*. Magnetic materials that are physically hard are usually magnetically hard, while those that are soft are generally magnetically soft too, but not always.

Magnetic poles   The lines of force emerge at or near the ends of the magnet. These are called the *poles* and any specially engineered terminations to direct or concentrate the field are described as *pole-pieces*. The poles are commonly termed

North and South respectively, but more accurately they are *north-seeking* and *south-seeking*. That is they are attracted to the earth's magnetic poles that are at its geographic north and south. As opposite poles attract, the earth's geographic north pole is the approximate location of its magnetic south pole. Furthermore, the magnetic and geographic poles do not coincide, which is why compass needles in England point to 16°W of true north, or nearer to NNW.

As a bar magnet is a string of molecular magnets end to end, the magnet can be broken across its centre to form two magnets, and further subdivided, each division producing magnets with a N and S pole. Two magnets can be joined end-to-end with a N and S pole in conjunction to become a single magnet. By dividing or joining, the pole strength is neither decreased or increased, the same number of lines of force run through the material.

Unit pole   A unit pole is defined as having the strength to exercise force of 1 dyne (10^{-3} newtons) on a similar pole at a distance of 1 cm.

Flux (ϕ)   The magnetic field consists of lines of force or flux each of which forms a continuous loop that passes externally from the N pole to the S, then internally back to the N pole. The c.g.s. unit is the *maxwell* which denotes one line of force; the MKS unit is the *weber* which is equivalent to 10^8 maxwells. One weber produces one volt when reduced to zero at a uniform rate, across a coil of one turn in one second.

Lines of force take the path of least magnetic resistance (*reluctance*), and so choose an external path through a material having the least reluctance. If there is no choice and the surrounding material is homogeneous such as air, they take the shortest path. However, they also repel each other and can never cross, so they balloon out from the magnet, each following the shortest path that does not encroach on another line. The stronger the field, the closer and more dense they are.

Field strength (H)   Field strength is rated in air (strictly a vacuum) being the number of lines of force in a given area. The c.g.s. unit is the *oersted*, which is one line of flux per cm^2. The MKS unit is the *ampere metre* (1 kA/m = 12.5 oersteds). It applies in particular to an *applied field* from a magnetic source to distinguish if from the *induced field* in a magnetic substance resulting from that field.

Flux density (B)   This is the density of flux induced into a material by an applied field. It is defined by the number of lines of force passing through a surface of given area perpendicular to the direction of the lines. The c.g.s. unit is the *gauss*, which is the flux produced by a field of one oersted, that is one line of flux per cm^2. The MKS unit is the *tesla* which is one weber per m^2.

For a field in air:

$$B = H$$

Permeability (μ)   An applied field induces a greater field than itself in a paramagnetic material. The magnitude of the induced field is a product of the magnetising force (applied field) and the permeability of the material. For diamagnetic substances the permeability is less than 1, for air it is 1, for paramagnetic materials it is greater than 1. For ferrous metals it can be several thousand. So:

$$\mu = \frac{B}{H} \quad \text{or} \quad B = \mu H$$

Reluctance   Magnetic resistance offered by a material having a length L, cross section area a and permeability μ:

$$r = \frac{L}{\mu a}$$

Saturation   A magnetising force causes the internal magnetic particles in a material to line up and so produce a coherent external field. The stronger the force, the greater the number of internal magnets that line up until finally all are so aligned. The material is then magnetically saturated: no further increase in force will induce an increase in flux.

Remanence/retentivity   When a magnetising force is removed, the induced field in a material drops. The remaining level is the remanence.

Coercivity (H_c)   This is the amount of applied reverse magnetic force needed to coerce the remaining magnetism of a material to zero after it has been magnetised to saturation. The material never ceases to be magnetic, rather the internal magnets are re-orientated so that there is no coherent external field.

Magnetic recording

Basic recording principles   Recording tape consists of magnetic material set in a binder on a flexible base that is capable of being magnetised in a series of individual, indivisible *domains*. As they pass the record head which generates a varying magnetic field corresponding to an analogue audio signal, each domain becomes magnetised according to the strength of the field existing at the moment of passing. When, in the replay mode, the tape again passes the head, the minute fields existing around each domain induces voltages into it which correspond to the original signal.

Initial magnetisation curve   The induced magnetic flux in any magnetisable material is not proportional to the applied magnetic force. As the force increases, the flux rises slowly to start with, then increases linearly to a point where it levels off, beyond which it reaches saturation. As a material can be magnetised from zero to either north or south flux values, there are two parts to the complete curve, one a mirror image of the other. If the magnetising force is in the form of an analogue signal such as a sine wave, with excursions either side of zero, the resulting flux is non-linear and the wave is distorted.

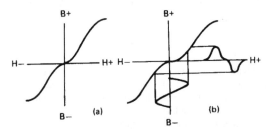

Figure 68. (a) initial magnetisation curve. *H* is magnetising force, *B* is the induced flux; (b) an applied sine wave produces severe distortion.

Hysteresis loop   (see Figure 69.) When demagnetising the material by reducing the force to zero, the flux diminishes slowly but remains at a substantial amount of its previous level (c). Reversing the polarity of the force brings the flux to zero (d), and if increased, builds the flux up in the negative direction to saturation (e). Reducing the force again to zero causes a slow drop in the flux (f), after which a further reversal of the force decreases it to zero (g), and subsequent force increase brings it up to the original point of saturation (b).

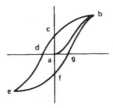

Figure 69. Hysteresis loop.

Bias

The non-linear portions of the initial magnetisation curve poses a problem for the recording of an audio signal. Any audio waveform recorded as an induced magnetic pattern is severely distorted as the applied magnetising force transverses the zero line unless some form of bias is applied. There are three methods of biasing.

D.C. bias from zero   The kinks in the curve can be avoided by adding d.c. to the audio signal through the record head. This magnetises the tape from zero to halfway up the initial magnetisation curve. Both half-cycles of the signal must thus be accommodated on one straight part of the curve, so recording amplitude is limited.

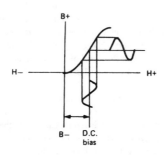

Figure 70. D.C. bias from zero.

D.C. bias from saturation   The tape is already saturated by d.c. through the erase head or a permanent erase magnet. The d.c. recording bias is of opposite polarity, so it reduces the flux from saturation. Thus the straight demagnetising part of the hysteresis loop is used instead of the initial magnetisation curve. Being longer it permits larger amplitude recordings. This is the method usually used where d.c. bias is employed.

The number of magnetic particles passing the head is constantly varying. With d.c. bias, these produce fluctuations of recorded flux. The result is noise and is the reason why d.c. bias is used only in cheaper recorders.

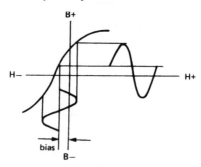

Figure 71. D.C. bias from saturation.

A.C. bias To avoid the problems of d.c. bias, the audio is superimposed on an a.c. waveform several times the value of the highest audio frequency. Theories differ as to how it works, but this is the most likely one. A tape domain follows several *hysteresis cycles* produced by the bias as it crosses the head gap. These cycles are smaller than the tape hysteresis loop because they are kept well short of saturation. When a positive-going audio signal is superimposed on the bias, the hysteresis cycles as a whole move up to a higher part of the loop and when the audio is negative-going they move downward to a lower one. So the mean flux level of the cycles moves up and down within the tape hysteresis loop in sympathy with the superimposed audio.

When a particular tape domain moves out of the main influence of the head gap the hysteresis cycle ceases at a certain point, and the domain is left magnetised to the level of some part of the cycle. As the mean flux levels of successive cycles are changing with the audio signal, the final magnetised level of each domain likewise changes. So varying states of magnetism remain that correspond to the audio waveform.

Fringe-field The field produced at the head gap is roughly hemispherical, so that which appears at the centre of the gap is longitudinal, while that at its edges is perpendicular. The needle-shaped tape particles are longitudinally orientated and are readily magnetised by the bias field over the central area of the gap. They acquire their final magnetisation level just before the edge when the field becomes perpendicular, which then

Figure 72. (a) bias produces hysteresis cycles within tape hysteresis loop; (b) superimposed audio displaces cycles vertically. A single point on the cycle thus traces the audio waveform. This remains when the domain leaves the gap.

exercises a force to change their magnetic orientation. However, long magnetic zones have a higher coercivity and remanence and so are more stable than short ones, They are thus affected little by the narrow perpendicular fringe field, but short ones are more readily re-orientated, So it is the short wavelengths corresponding to the high frequencies that suffer partial erasure. This is one of the several limiting factors to the upper frequency response.

Bias superimposition The audio is not modulated on to the bias waveform, but superimposed: an important distinction. The bias carries it on positive and negative half-cycles without its own amplitude being changed. With modulation, the amplitude of the high frequency carrier varies according to the modulation, and sidebands of plus and minus the two frequencies are produced. At low bias frequencies these can produce spurious tones in the audio range. If the bias gets into the recording amplifier, such intermodulation can occur, so filters are often introduced to block it and bias frequency is kept as high as the head will allow.

Optimum bias Too low a bias restricts the straight portions of the hysteresis cycles giving non-linear audio recording. If too high, the bias saturates the tape also producing distortion. There is then an optimum position for minimum distortion for any given tape.

Too high a bias also increases the fringe field so reducing h.f. response. The optimum points for minimum distortion and maximum h.f. response do not coincide. A compromise must therefore be chosen between the two, or the setting can be made to favour one or the other. Some tapes are less critical than others, so are compatible with many different machines. Others, such as chrome, need a specially high bias setting.

118

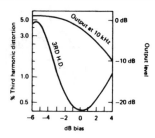

Figure 73. Typical tape curve plotting distortion and h.f. response.

Crossfield bias To avoid loss of h.f. due to the gap-edge field, an auxiliary magnetic field is set up between the head and a pole piece placed behind the tape. At the leading edge of the gap the auxiliary and the perpendicular gap-field are in the same direction and reinforce each other, but at the lagging edge they are in opposition and virtually cancel. Gap flux density is thus asymmetrical with very little perpendicular field appearing at the lagging edge to affect the recorded short wavelengths.

Another method uses a separate head for bias which is positioned behind the tape and offset from the audio head gap. The tape undergoes hysteresis cycles as it passes the bias head while also within the field from the audio head, so an audio signal is recorded by the combined fields. Being offset, the bias field is asymmetrical and so has little effect on the audio leaving the lagging edge of the record head.

Crossfield bias can improve the response of open-reel recorders but would be difficult to implement with cassette machines. The improved h.f. response resulting from tape and head developments has lessened its appeal and it has fallen from favour for domestic machines.

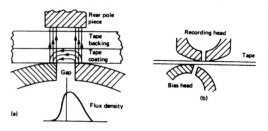

Figure 74. Crossfield system with (a) auxiliary field; (b) separate bias head.

Record/playback heads

A single head is usual for both functions but the requirements
are different so single heads are a compromise. More expensive
machines have two heads; this enables *A B monitoring* (hearing
the playback off the tape while recording) to be performed,
permits the optimum impedance to be used for each head and
avoids head switching. Where space is limited such as in
cassette decks, the two heads may be accommodated within the
same casing.

Construction   Each unit consists of usually two coils wound
on a ring-shaped magnetic core, which is normally squared off
at the rear end. The two coils being on opposite sides of the
core are connected in opposite phase; this makes them less
vulnerable to external hum fields. At the front there is a gap
that is filled with non-magnetic shim, across which the
magnetic fields are generated. With stereo heads, two units are
mounted one above the other. The proximity and identical
orientation of the two sets of coils make coupling and
consequent crosstalk between channels a strong possibility.
Careful magnetic screening must be employed to minimise
this.

Playback gap   When a recorded wavelength on tape equals
the width of the gap, both half cycles being within the gap at
the same time cancel and there is zero output. Output increases
as the wavelength lengthens to a maximum at twice the gap
width. Thus the gap width is one of the limiting factors for the
high frequency response.

   The tape speed for a cassette is $1\,^7/_8$ i/s or 47,500 μm/s. So
to determine the maximum gap (g) in microns to read a
particular frequency (f):

$$g = \frac{47,500}{2f}$$

Hence for a response up to 20 kHz, the gap width should be no
greater than 1.2 μm.

Record gap   Record-only heads frequently have gaps up to
ten times the width of playback heads. They may thus seem
incapable of recording short wavelengths. However, as a
section of tape crosses the record gap, it is magnetised through
many hysteresis bias cycles until it leaves at the lagging edge
with a particular magnetic value imposed by the level of the
audio signal at that instant. It is thus at the lagging edge that the
final recording is made, so gap width does not affect recorded
frequency.

The field produced by the gap is roughly hemispherical and its radius is proportional to the gap width. Depth of penetration into the coating is approximately equal to the field radius, so it follows that a wider gap results in deeper penetration. This means that more of the magnetic material is utilised, so a stronger signal is recorded and the signal/noise ratio is improved.

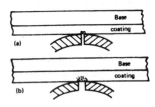

Figure 75. Field penetration with (a) a narrow gap; (b) wide gap.

Frequency response and losses The magnitude of the signal induced into the playback head is proportional to the velocity of the flux passing the head gap. As velocity decreases as the frequency drops, the response falls toward the bass at the rate of 6 dB/octave. There are several losses at the high-frequency end.

Self-erasure The small magnetic particles in the tape line up to form larger magnets laid end-to-end, one for each half cycle of the recorded signal. As successive half cycles are of opposite polarity, the magnetic poles become: N–SS–NN–SS–N, with like poles adjacent. This arrangement is conducive to self-demagnetisation, but short, thick magnets are more liable to it than long narrow ones. The short wavelengths thus suffer the most, resulting in high frequency loss.

Head gap The effective playback head gap limits the h.f. response. Wear of the pole faces, oxide deposit lifting the tape from the head and badly aligned head laminations, can all make the effective gap larger than the nominal size.

Fringe field The influence of the perpendicular bias field on tape at the lagging edge of gap erases the shorter wavelengths.

Head losses Hysteresis losses occur in the head core which are greater at high frequencies than low.

These various h.f. losses together with the falling bass, result in a frequency response having a large peak at about 3 kHz. This must be corrected if an overall flat response is to be obtained.

Pre-emphasis A degree of high frequency boost is applied to the recording amplifier, which also helps the signal/noise ratio as most of the noise energy is at high frequencies. Too much pre-emphasis could cause tape saturation, so it must be limited. Different amounts are applied for chrome and ferric tape.

Equalisation Applied to the playback amplifier, equalisation boosts frequencies below 3 kHz by 6 dB/octave. The high frequencies are also boosted by an amount that is specified as a *time constant*. This is the time a capacitor takes to charge through a resistor to 63% of its maximum value and is a convenient way to describe the combination needed to produce the required response.

For ferric tape, the equalisation is 120 μs, which lifts frequencies above 1.2 kHz, and for chrome it is 70 μs which starts at 2.2 kHz. Because of the influence of the bass boost, the curve does not flatten and start to climb until 3 kHz and 4 kHz respectively.

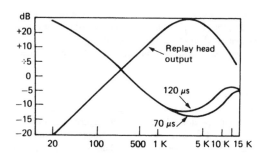

Figure 76. Characteristic and equalisation curves.

Noise reduction

Tape recorded noise has three principal causes apart from noise in the record or playback amplifiers. These are:

Non-uniform magnetic particles   These differ in size and shape, in their crystalline structure and in their interaction with adjacent particles. Their coercivity and remanence thus differs, so each is left in a slightly different magnetic state of orientation than its neighbour. This produces a varying flux as they pass the head gap, resulting in noise.

Modulation noise   As the signal waveform rises and falls, different numbers of magnetic particles are magnetised out of the total leaving the gap at any instant. The effect is noise, which is greater with large waveform excursions. It is particularly noticeable with heavy bass signals which do not mask the high noise frequencies as effectively as the treble signals.

Asperity noise   Unevenness of the coating surface produces displacements of the tape as passing asperities lift it from the head. These cause random signal variations, hence amplitude modulation of the signal. The sidebands of this modulation are heard as noise.

Many noise reduction systems have been devised; most are based on the principle of the compander and on the facts that tape-generated noise consists mainly of high frequencies, and that loud sounds will mask quieter ones in the same frequency band.

Compander The recording unit compresses the range between loud and soft signals (dynamic range) so that the softer ones are made proportionally louder. If an original range of 80 dB is compressed to 40 dB, there is a 2:1 compression, but because this is a linear expression of a logarithmic ratio, the term is said to be *log/linear*. The actual reduction would be from 10,000:1 to 100:1 in this case. The playback unit expands the range to its original extent by reducing the quieter signals. In doing this the tape noise is also reduced by a similar proportion. Though seemingly an ideal solution, simple companders suffer from several disadvantages.

Noise pumping   As playback gain is controlled by the signal level, the remaining noise varies with the signal. Although it is masked at high levels, it can be heard to drop after loud sounds cease due to the time taken for the gain to re-adjust. Variations can also be heard during low-frequency sounds because of their poor masking effect.

Inaccurate tracking   If the expander does not follow the compressor characteristics exactly, amplitude distortion and dynamic errors occur. Gain control of both sections is derived from rectifiers that sample and process the signal; these tend to be non-linear and do not respond fast enough. So matching between the two units is not easy to achieve.

Frequency distortion exaggeration   A low and a high frequency tone of equal amplitude are fed consecutively into a recorder having response deficiencies that cause one tone to be recorded at a lower level than the other. On replay, the expander senses its lower level, and reduces it still further; thus recorder frequency-response deficiencies are exaggerated. This does not happen when the two tones are recorded simultaneously as there is then just one programme envelope to control the expander gain.

dBx   This system uses a compander, but minimises the pumping effect by pre-emphasising the high frequencies which rise from 200 Hz to a level of +12 dB at 5 kHz and above. This would lead to tape saturation problems, but that is avoided by applying pre-emphasis also to the signal controlling the compression. The amount is +20 dB at 1.6 kHz, so by increasing the compression at that frequency, the actual recorded response is –5 dB at 15 kHz. By similar de-emphasis in the expander signal and control paths, the response is equalised.

The control signal is obtained from special r.m.s. detectors that sample an average unit of programme area instead of measuring its instantaneous peak value. It is thus unaffected by sharp spikes, drop-outs, or phase errors and gives accurate control that can be matched by the expander.

Recorder frequency deficiencies suffer exaggeration as with the simple compander, but this is not a major problem if the system is used with good quality, well-maintained equipment that has a flat response. It offers 30 dB noise reduction at all frequencies.

High Com   Developed by Telefunken, this system is basically a compander. The larger the noise reduction ratio achieved by a compander, the greater also are its undesirable side effects. High Com gives more modest noise reduction, with a corresponding lessening of the disadvantages. As tape noise consists mostly of higher frequencies, the reduction is 10 dB below 1 kHz, rising to 25 dB at 10 kHz.

The control signal detectors have a very fast attack time and are combined with special overshoot-limiting circuitry. The result is a high degree of accuracy in matching compression with expansion.

The system is not prone to error due to different output levels from tapes of different sensitivities, a problem with some other noise reduction systems. Also it seems not to suffer from recorder frequency-response exaggeration. It is thus well suited for domestic use where conditions are not always ideal.

ANRS The Automatic Noise Reduction System from JVC uses a compander that is frequency selective. High frequencies are boosted by an amount that decreases with their level. The –4 dB level is boosted by a maximum of 10 dB at 5 kHz, the –30 dB level by 8 dB at 3 kHz, the –20 dB by 5 dB at 1 kHz, while the –10 dB level is boosted by 2 dB at 500 Hz. At all levels the boost starts at 200 Hz. This is achieved by passing the control signal through a high-pass filter to make it frequency selective, rectifying and smoothing it, then using it to control a variable filter in the main signal path.

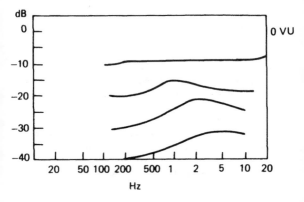

Figure 77. ANRS encoding frequency response.

The filter is a bridge circuit having a capacitor across it and a transistor used as a variable resistor in one leg. When it is balanced, the capacitor has no effect, but when unbalanced by the control signal applied to the transistor, the capacitor shunts the bridge and increases its high-frequency response. To avoid intermodulation distortion arising from a.c. components in the control signal, its time constant varied according to signal level from 5 ms at low to 2 ms at high levels.

During playback the same circuit elements are used but the variable filter appears in a negative feedback circuit, which modifies the playback response to the inverse of that of the recording amplifier.

Dolby B   This most common domestic system is used under licence by a large number of tape deck makers. It is a compander in which the signal is split into two paths, the *main* and the *sidechain*. In the main path, the signal is unchanged during recording and playback: but in the sidechain high frequency signals of a –40 dB level are boosted 10 dB at 5 kHz; those of –30 dB, by 9 dB at 2.5 kHz; levels of –20 dB by 5 dB at 1.5 kHz; and those at –10 dB by 0.5 dB at 1.5 kHz. Boost starts at 200 Hz in each case.

The 0 dB reference level for the above values is a recording flux of 200 nanowebers/metre. Thus the maximum noise reduction is of 10 dB above 5 kHz.

These values are similar to ANRS, but the principal difference is the split paths. After processing, the sidechain is added to the main one so producing a boost of those high frequencies in the overall recording response. In the playback mode, the sidechain performs the same process, but is subtracted from the main path thus levelling the combined record/playback response.

The use of sidechains means that the bulk of the programme material passes through the main path and so is not processed. Hence the frequency error exaggeration and loss of transients that can occur with companders is confined to the small part of the total signal in the sidechain.

Noise pumping is also minimised for the same reason. It is to avoid the possibility of pumping by high-level mid-frequencies that the threshold frequency increases as the level rises.

Dolby C   This improved version was introduced to match the general improved performance of cassette decks and tapes. The amount of boost is greater than with B; the curves start lower at 50 Hz and level off at a lower frequency.

The −60 dB level is boosted by 20 dB at 3.5 kHz, the 40 dB level by 16 dB at 1 kHz, the −30 dB by 13 dB at 800 Hz, the −20 dB level by 9 dB at 600 Hz and the −10 dB by 4 dB with a very gentle slope. Noise is thus curtailed right through the mid-frequency range, and is at a maximum of 20 dB for the lowest level, above 3.5 kHz.

These boost values could produce tape saturation at high frequencies. To avoid this the curves fall back gently above 2 kHz and sharply above 10 kHz. Most recent decks having the B system also have C.

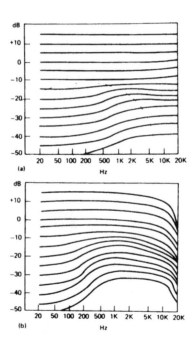

Figure 78. Dolby encoding frequency responses: (a) B system; (b) C system.

Dolby A   A professional version having four sidechains covering the frequencies low–80 Hz, 50 Hz–3 kHz, 3 kHz–high and 9 kHz–high. The two overlapping high bands give greater noise reduction where it is most needed. Band-splitting gives several advantages.

Only low frequency material compresses low frequency bands, so modulation noise is almost eliminated. The time constants for each band can be optimised and compressor/expander tracking is less critical because an individual band mistracking affects a smaller portion of the complete signal.

With all Dolby systems, though, recording *and* playback levels need to be set correctly to avoid anomalies in the frequency response.

DNL   The dynamic noise limiter is a completely different system by Philips that works on playback only. It uses the principle that when music is played softly the instruments generate few harmonics above 4 kHz: but most noise frequencies are higher than that. Loud music has many more harmonics but masks the noise. So the system cuts frequencies above 4 kHz severely when the signal level is low and with it much of the noise, but reduces the amount of cut as the level rises.

It does this by splitting the signal into two paths which are in opposite phase. In one the signal is unmodified but in the second, it passes through a high-pass filter which removes everything below 4 kHz, then through a variable attenuator which is controlled by the signal level.

Finally, the two chains are added, but as they are in opposite phase the remaining high frequencies in the second path cancel those in the first, to an extent governed by the variable attenuator. As there are no low frequencies in the second path these are not cancelled.

Noise is reduced by some 10 dB at 6 kHz, and 20 dB at 10 kHz and as the process works only during playback any recording can be so improved. Noise below 4 kHz, though less objectionable, remains and some harmonics above that point could be lost. It is therefore a compromise system, although a good one.

Head circuits

Head impedance   Separate heads are connected directly to their respective amplifiers, the record head to the output of the recording amplifier and also to the bias oscillator, while the playback head is connected to the input of the playback amplifier. Impedances can thus be optimised. Magnetic flux is proportional to current flowing in the head coils and is not affected by voltage. So low impedance is needed for the record head to permit the required current to flow: it also best matches the recording amplifier output.

For playback, the recorded flux produces an e.m.f. (voltage) in the head coils. To present a high signal voltage to the input of the playback amplifier and thus minimise amplifier noise, the head coils should have a large number of turns and so have a high impedance compared to that of the record head. For single record/playback heads, the impedance must be a compromise.

Switching   Single heads must be switched to their respective circuits, and the usual way of doing this is by means of a two-way, single-pole switch for each channel. If the 'live' terminal of the head was switched, the switch and its wiring could pick up hum and noise. Instead. the connections to the replay circuit are permanently taken to one side of the head, while those to the record circuit are connected to the other. The switch connects each side to earth in turn, so the switch is at earth potential and the amount of live signal wiring reduced.

A further switch may be used to switch either the playback head or a microphone into the input of the amplifier. The erase head is not usually switched because the oscillator is switched off during playback.

Recording frequency correction   A recording head consists of inductance and resistance in series, the resistance being the d.c. value of the coils. The impedance is equal to:

$$Z = \sqrt{R^2 + X_L^2}$$

where Z is the impedance; R is the resistance: and X_L is the reactance. At low frequencies X_L is small compared to R, but at high frequencies it is the opposite. So at high frequencies progressively smaller currents would flow resulting in a poor high frequency response and signal/noise ratio. The phase angle also changes with the ratio between R and X_L so phase errors would also occur with changing frequency.

The most common solution is to increase the value of R by a large amount so that it swamps the changes in X_L and ensures that the circuit is mainly resistive. This is done by wiring a resistor in series between the head and the record amplifier.

In addition a capacitor is connected across the resistor; this increases the high frequencies and further improves the phase response. The value of these components depends on the values of R and X_L for the particular head to achieve optimum overall response. This may be upset if a replacement head has widely different R and X_L values.

A large amount of recording signal is lost in the series R/C circuit, so the recording amplifier must be able to make up this loss.

Recording power   In order to drive the required amount of current through the large value series resistor and the record head, a high signal voltage is also needed from the recording amplifier output. It must therefore provide power and so needs to have a power output stage. In portable recorders with a single switched amplifier. the recording output is often taken from the loudspeaker output stage.

Slew rate   With power amplifiers there is a limit to the rate of change that the output voltage can follow. At high frequencies the rate of change is greatest, but this depends on the signal level. At high levels the rate of change is greater than at low levels. The ability of an amplifier to follow large rates of change is termed its *slew* rate. With a recording amplifier where high voltages are required, this sets a limit on the h.f. performance attainable and also can produce intermodulation distortion.

Transistor head drive   To avoid the use of a high-value series resistor, in some models the head is driven by a transistor stage that incorporates frequency correction, or a *transconductance converter*. An example of the latter is the *actilinear circuit* which uses a pair of complementary push-pull transistors with negative feedback which supplies constant current to the record head irrespective of the variation of its reactance with frequency.

Head drive adjustment   For the measurement of recording level, a 10 Ω or 100 Ω resistor is often included in the earth end of the record head circuit. A sensitive millivoltmeter is connected across the resistor and a fixed tone is injected into the microphone or auxiliary socket at a level to give a 0 dB on the VU meter. The recording level preset is then adjusted to give the reading specified in the service manual. This varies with head impedance. but a typical value over a 10 Ω resistor is 0.35 mV for ferric tape and 0.7 mV for chrome.

Bias level adjustment   The bias does not pass through the high-value swamping resistor but is fed to the record head through a variable resistor, capacitor, or tunable coil with which the level is set. It is measured across the 10 Ω or 100 Ω series resistor. Values vary, but are roughly ten times the maximum recording level, ranging from 3.0–7.5 mV across 10 Ω or 30–75 mV across 100 Ω. Some but not all decks switch to a higher value for chrome tape.

If the maker's figures are not available or a different type of tape is to be used, the bias can be set by recording a 400 Hz and 10 kHz tone at the same low recording level, say 12 dB, then comparing the levels at playback. If the 10 kHz tone is lower, the bias must be decreased and another recording and replay comparison made. Repeat until either the two levels are the same for optimum h.f. response, or the 1 kHz tone is 1 dB down for optimum low distortion.

ALC   Automatic level control is universal on portable recorders. Part of the output of the record amplifier is rectified and used to control an earlier stage. Time constants of the circuit are chosen to give a rapid attack but a slow decay. Sudden loud signals are thereby prevented from overloading the tape, while the gradual fade-up prevents level changes being too obvious. A general level is maintained by the average signal.

Serious users find ALC more of a disadvantage than a help because level changes are noticeable, especially when there are background noises. Early cassette recorders were fitted with a manual/auto switch, but these no longer appear. Some owners modify their recorders to cut out the ALC.

Head materials

Permalloy   The standard material is Permalloy, an alloy of
about 78% nickel and iron with some molybdenum. It accepts a
high flux density without saturating, so it will take high
recording levels without distortion. The permeability (magnetic
conductivity) drops as the frequency rises being only a tenth of
its 1 kHz value at 10 kHz, so the h.f. performance is poor.

Eddy currents circulate within the core because Permalloy
is a good electrical conductor. These increase with frequency
and produce losses in the form of heat. They further impair the
h.f. response and limit the maximum usable bias frequency. A
low bias frequency increases the possibility of intermodulation
distortion.

The core is laminated to reduce eddy currents but this
creates its own problems. It is impossible to align the stack so
as to give a perfect straight-sided gap. Deviations either side of
the vertical enlarge the effective gap width, and impairs the h.f.
response when used for playback. With gaps of the order of
1 μm, microscopic deviations are significant.

Permalloy rates 130–140 on the Vickers hardness scale
which is the softest for record head material and wears rapidly.
Average rate of wear is 120 μm per thousand hours. Average
life is 1000–2000 hours, but this depends on the tape used and
whether the tape is in head contact during fast winding for
programme searching. The non-magnetic shim to fill the gap is
chosen to have a similar rate of wear as the core material; for
Permalloy, beryllium copper is usually used.

Sintered ferrite   This is a combination of various oxides,
mainly iron oxide, zinc oxide, manganese oxide and nickel
oxide, in fine grain form with a ceramic filler and binder. It
permits half the maximum flux density of Permalloy and a
tenth of its permeability, but this is less dependent on
frequency, being three-quarters of its 1 kHz value at 10 kHz.

Ferrite has a high electrical resistance, so eddy current
losses are minimal and higher bias frequencies can be used. It
can thus be made from a solid block instead of laminations and
so can be given a more accurate gap. However, the material is
brittle and chips easily, also tiny bubbles can form in it during
manufacture. These can result in cavities when the head face is
machined into shape. The residual magnetism is greater than
Permalloy and so it requires a larger coercive force to
overcome it, possibly resulting in a higher noise level.

The hardness is 400 on the Vickers scale, which is three times that of Permalloy. So while having superior h.f. response and hardness, it does not permit high recording levels and can result in higher noise.

HPF   This stands for hot-pressed ferrite which is made by compressing ferrite at pressures of the order of 7000 lb per square inch (48 MPa), at temperatures around 1400°C. Permeability is greater than sintered ferrite and even that of Permalloy, while its dependence on frequency is a half of the 1 kHz value at 10 kHz. Residual magnetism is less than ordinary ferrite and Permalloy. No bubbles can remain in HPF because of the high pressure so the possibility of unexpected cavities is eliminated. It can take a high polish thereby minimising tape friction, and has a Vickers hardness rating of 650–700. Hard glass is used for the gap filler. Wear is around 0.4 μm per thousand hours. Maximum flux density is the same as ferrite, which is half that of Permalloy.

Sendust   The qualities of HPF make it ideal for all applications except metal tape which requires high flux densities to record and erase. Sendust has a similar flux sensitivity to Permalloy but with the hardness of ferrite. Its disadvantage is its low electrical resistance which leads to eddy currents and h.f. losses. This has been overcome by rolling it into ribbon when hot, then making laminates from the ribbon. However, laminates produce staggered stacks and erratic gap widths.

Head overload   As the flux density in the head core increases, so does the distortion level. As with tape, this is primarily third harmonic distortion, which becomes severe as the flux approaches saturation. It is not a problem with Permalloy heads and ferric tapes as tape saturation occurs long before that of the head, but iron tape and to a lesser extent, chrome, need a high recording level which can drive ferrite cores well toward saturation and so into high distortion levels.

Heavily pre-recorded tapes can produce high levels of distortion by driving the replay head toward saturation.

The search for the ideal material continues but one solution is to use Sendust for record and erase heads where high flux densities are required but gap width is unimportant and HPF for the playback head where gap uniformity is a major requirement.

Table of head materials and their properties

	Permalloy	Ferrite	HPF
Composition	Ni	MnO	MnO
	Fe	Fe_2O_3	Fe_2O_3
	MO	ZnO	ZnO
		NiO	NiO
Permeability			
1 k	18 000	1200	20 000
10 kHz	1500	900	10 000
Maximum flux density (tesla)	0.7	0.4	0.4
Coercive force* (oersted)	0.02	0.5	0.015
Specific resistance (Ω/cm^3)	0.000 05	>100	>100
Vickers hardness	135	400	700

*to overcome residual magnetism.

Erasing

All tape recorders erase unwanted recordings as new ones are made. This is accomplished by the erase head over which the tape runs first. The record head is next and playback head last if there are separate heads for these functions.

With machines that use d.c. bias, erasing is accomplished by either passing a large d.c. current through the erase head or using a permanent magnet. In the latter case, the magnet must be retracted from the tape path during replay. Sometimes a magnet is used in recorders having a.c. bias.

The full output from the bias oscillator is used for a.c. erasing, so its value is not usually preset, but the erase current is generally in the order of 100 times that of the bias current. The erase head has a large gap (typically 200 μm), which covers almost half the width of the tape so erasing both mono and stereo tracks. Some heads have an E-shaped core having two gaps to provide a double erasing action.

Effect of oxide deposits

Loose oxide particles are deposited on the head and others become attracted to them. In time they build up to a blob of significant size which is compacted by tape pressure and resists removal. It prevents intimate tape/head contact at the site and lifts the tape on either side in a 'tenting' effect. Thus quite a small deposit that is hardly visible can affect performance considerably.

Effect on recording   Some of the flux passes in the space between the tape and head, so less penetrates the tape coating. If the blob is across the gap it will also form a magnetic shunt to further reduce the flux passing through the coating. So, the tape receives a low recording level and the signal/noise ratio deteriorates. The effect is not particularly frequency dependent.

Effect on playback   Less of the recorded flux is applied across the gap so total playback level is reduced. The side field from a magnet extends only half as far as one that is four times its length. Thus fields from short recorded wavelengths extend only half the distance from the tape surface, as those two octaves below. So when the tape is lifted from the gap by a deposit, less flux from the shorter wavelengths reach the gap than that from the long ones. The result is loss of high frequencies.

A tape that is both recorded and played with a dirty head will sound muffled, but may be acceptable when the head is cleaned. This is because the h.f. loss occurs during replay; a recording made with a dirty head will be at a lower level and have more noise, but may suffer little h.f. impairment.

Flutter   The deposits tend to be sticky and so may cause the tape to stick and snatch resulting in the rapid speed fluctuations known as flutter. If this occurs while recording, the effect is permanent, but if only during playback, will disappear when the head is cleaned.

Head adjustments/maintenance

Cleaning   Cleaning cassettes consist of a length of woven material that removes deposits by friction. Over-use can cause excessive head wear. Once every 25 hours is recommended. Discard the cassette when woven tape appears soiled otherwise oxide may be deposited on the head being cleaned. More elaborate cleaning cassettes apply an oscillatory rubbing action driven mechanically from the take-up hub. These usually need charging with cleaning fluid.

The best method of head cleaning is by hand. If done with care it is more thorough than a cleaning cassette and causes no wear. It may be necessary to put the equipment in the 'play' mode to afford easier access to the heads, but without any power applied. A cranked handle dental-type mirror is invaluable for inspection. A cleaning bud consisting of a pad of cotton-wool on a stick soaked with industrial alcohol, special cleaning fluid, or methylated spirit should be used on the head until all trace of oxide has gone. While the head is still moist from the fluid, polish off with a dry, clean bud. This is essential to remove any sticky residue from the fluid. Do not touch the head after cleaning.

The rubber pinch wheel should be cleaned in similar manner. Deposits here can cause speed irregularities.

Wear   A flat channel the width of the tape is gradually cut across the head face. In itself this has little effect, but if uneven, the tape can lose intimate contact. Uneven wear around the gap extends its effective width and so reduces playback h.f. responses. Developing crevices between laminations or in the gap due to break-up of the shim, can attract oxide particles which rapidly build up deposits. Worn heads are thus likely to need cleaning more frequently than unworn ones.

Examination under high magnification can reveal these defects. If not present and if playback volume and frequency response tests are satisfactory, the head can be considered to have further useful life in spite of a worn channel.

Magnetisation   In time the head can become permanently magnetised with the effect of adding noise to recordings. Demagnetisation is therefore advisable from time to time. A magnetic probe, which develops a strong a.c. field at mains frequency, is introduced to the head and slowly moved away. The head material thereby goes through successive and decreasing hysteresis cycles until at removal the residual magnetism is zero. Some head-cleaning cassettes also do this automatically by means of internal revolving magnets, but the effect is less positive.

Azimuth   When the record head is vertical, the gap records vertical magnetic zones along the tape. If a playback head is not vertical, its gap forms a diagonal across the track and so can bridge adjacent zones of short wavelength. Its effective width is thereby increased and h.f. response is reduced.

If the record head is not vertical, the recorded zones are diagonal and a vertical playback gap bridges them, again causing loss of h.f. If however, a non-vertical head both records and plays back, there will be no loss because the replay gap is aligned with the zones on the tape. Hence tapes recorded on a machine with a mis-set head are satisfactory when played on the same machine but not on a properly adjusted recorder.

With stereo machines a non-vertical replay head means that one gap is slightly ahead of the other which produces spurious h.f. phase differences between channels, hence poor stereo location.

Azimuth adjustment   The head can be rocked from one side to the other by means of a set screw at one side, and a compression spring under the head bracket holds it in the set position. Adjustment can be made by playing a tape recorded with a 10 kHz tone on a good machine and adjusting for maximum output measured with an output meter.

For stereo machines, the output of each channel can be fed to the X and Y inputs of an oscilloscope and a lissajous figure obtained from the 10 kHz recorded tone. Azimuth is adjusted for zero phase difference.

If no test equipment is available, play a recording of some white noise (inter-station hiss on f.m.) made on a good machine. Adjust azimuth for maximum treble. Swing adjustment either side of centre and gradually decrease excursions. With care and a keen ear an accurate setting can be achieved.

Head height and fore-and-aft tilt must also be correct for the tape to follow a perfectly straight path. This is pre-set with most cassette recorders but adjustment may be required for open-reel machines.

Tape transport

The transport system must pull the tape off the supply spool, pass it over the heads at a constant speed which must not deviate over its whole length, and then evenly wind it up on the take-up spool. It must also be able to be re-wound at a fast speed.

Main drive A belt drive from a pulley on the motor shaft to the flywheel rim is the almost universal method. The belt also drives the take-up hub, or this may be driven by a secondary belt or intermediate rubber wheel from the flywheel.

Tape drive The flywheel spindle is also the one that drives the tape. It must be machined to within very fine limits otherwise cyclic speed variations will occur. To give a positive drive it is often left unpolished. The tape is pinched between it and a rubber roller. This overlaps the tape on both sides and is therefore driven by direct contact with the spindle. The area of tape in contact with the roller is greater than that in contact with the spindle because of its larger circumference, so the principal tape traction comes from the roller rather than the spindle. Its surface must therefore be perfectly even and its bearings free. As the traction exerted by the pinch is limited it is essential that the paying out spool is free from excessive or uneven friction. Sometimes moderate friction is applied to brake the spool and prevent it unwinding loose tape.

Tape take-up The speed of the take-up spool varies continuously throughout the length of the tape. This is because the amount of tape taken up by each revolution increases with the circumference of the tape on the spool. So as it fills it slows. Drive to the take-up hub cannot therefore be direct, but must be via a *slipping clutch*. It consists of two plates separated by a felt pad, held in contact by gravity or an expansion spring. Torque is transmitted through the friction of the pad. It must be sufficient to overcome a possible excess of hub friction, but not too great so as to heavily load the drive system when the spool is turning slowly and slow it down.

To avoid these problems. better decks use a separate motor to drive the take-up spool. It is under-run to reduce its torque to avoid straining the tape and influencing its speed through the pinch drive.

Fast wind/rewind   For fast wind the clutch is bypassed, often by introducing an intermediate wheel to drive the upper clutch plate from the flywheel rim. At the same time the pinch roller is released to allow the tape free passage. With machines having a separate motor, it is switched to full power to obtain maximum torque. Rewind is obtained in a similar manner, usually by intermediate wheel drive from the flywheel to the paying out hub. Decks having a separate motor for winding often have a third one for rewind.

Pause control   Portable recorders often have a pause control comprising a switch in series with the power supply. This economises in battery power but can record noise on starting. Furthermore a permanent crease can be impressed in the tape if it is left pinched against the spindle for more than a few minutes, so this type of pause control has little to commend it. The preferred mechanical control brakes both take-up and supply spools while lifting the pinch roller from the spindle. The flywheel continues to be driven from the motor, so there is an instant start when the pause control is released.

Pressure pads   Intimate contact between head and tape is normally maintained by means of a pressure pad. Each cassette has its own, but with open-reel machines it is part of the recorder. Pads bearing an oxide deposit can cause flutter, while those displaced or poorly sprung can result in drop-out (momentary signal cessation) and loss of h.f.

These effects can be avoided by maintaining tape tension across the heads by other means. This is imperative with three-head cassette decks as there is only a single pressure pad in each cassette. One method is to drive the paying-out spool backwards through a slipping clutch or low-torque motor, The disadvantage is that the mechanical load increases as the tape proceeds just as that of the take-up spool does. Such a double increase in load could slow the tape throughout its length.

Dual-capstan drive   Back tension is here achieved by having a second capstan and pinch roller drive before the heads and driving it some 1% slower than the main drive. This is done by having identical spindles and flywheels driven by the same belt. As the motor pulls the belt from flywheel A the tension stretches it slightly, but feeding it to flywheel B reduces normal 'at rest' tension so producing a contraction of that portion. So A is driven by a stretched belt portion, and B by a contracted one, This affects the relative gear ratios with flywheel A revolving some 1% faster then flywheel B.

If the rotation direction is reversed, flywheel B becomes the faster, hence this arrangement can accommodate tape reversal if the heads are made to read the bottom tracks. Direction reversal though means the flywheels must come to rest and restart in the opposite direction, so reversal is not instantaneous.

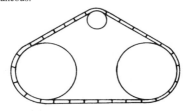

Figure 79. Varying degree of stretch of belt driving two flywheels.

Reversing systems   Mechanical details differ, but these are the basic principles. Two spindles rotate simultaneously in opposite directions The pinch rollers disengage the one and engage the other as reversal is initiated, so the changeover is rapid.

The heads can be four-track or two twin-track heads in the same case, with one pair switched in for each direction. The left-hand channel must always be switched to the outer track Another arrangement rotates the record and replay heads through 180° as one unit, thereby preserving the correct sequence and track positioning for the reverse run. Stainless steel stops and ruby-tipped stop screws ensure accurate rotation angles and azimuth settings that are unaffected by wear.

The erase head must always be in advance of the record head, so for tape reversal two erase heads are needed. Normally it occupies the unused spindle drive cassette aperture, but this space is now taken by the reverse drive spindle, so the erase heads must be accommodated in the two small apertures either side of the centre one.

Motors and speed control

Mains driven Mains recorder motors are of the *synchronous* type as these are controlled by the mains frequency and run to a constant speed. The rotor is a permanent magnet which follows a rotating field set up by the stator windings. This is achieved by splitting the phase to alternate poles, or by shading one pole with a copper ring. Two-pole motors run at 3000 r.p.m. and four-pole at 1500 r.p.m. at 50 Hz. Bearings are usually made of sintered bronze (powdered metal compressed under heat) which acquire lubricant from a surrounding felt pad by capillary attraction. Being of spherical shape and retained by a spring clamp, they are self-aligning; any stiffness of the rotor spindle can usually be cured by a slight tap.

Mechanical governor Early battery recorders used a d.c. motor having a short springy arm on the rotor bearing a small weight and a contact that mated with another on a rigid arm. A set-screw varied its tension and position. The contacts were normally closed, but when the rotor reached a set speed, centrifugal force parted them interrupting the current through the windings. The motor slowed and the contacts re-made. This happened several times a second, the fluctuating speed being smoothed out by the flywheel. Speed adjustment was on a trial and error basis by means of the set-screw accessible through a hole in the motor casing.

Electronic governor The circuit shown in Figure 80 is widely used, with minor variations, in most cassette recorders. It compensates for falling battery voltage as well as load variations. The current through the motor is controlled by the series transistor Tr2 for which the forward bias is supplied through Tr1 and R4. Bias for Tr1 is derived from R1, R2, and R3 connected from the supply to the collector of Tr2. A drop in the supply decreases the voltage across the motor and R4 which is virtually across it. Being the emitter resistor for Tr1, the reduced voltage increases current through Tr1. This in turn biases Tr2 harder on, thus increasing the motor current, so compensating for the reduced supply voltage.

A load increase slows the motor reducing its back e.m.f. and increasing its current which also flows through R5. The voltage drop across it is increased thereby increasing the forward bias and current through Tr1. Tr2 is thus driven harder, further increasing the current through the motor. Greater torque is thus produced to overcome the extra load.

When switching on, both transistors are without forward
bias so the motor would not start. R6 applies a voltage to the
base of Tr1 to get things going but has little effect
subsequently. The diodes compensate for temperature changes
and R2 is variable for speed adjustment.

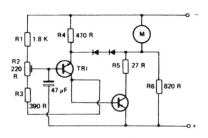

Figure 80. Widely-used electronic governor circuit.

A.C. feedback control   The voltage or current-change
sensing circuit reacts at moderate speed and produces a hunting
effect. A.C. feedback control responds faster, thereby reducing
hunting and giving the flywheel smaller speed fluctuations to
smooth. Furthermore, supply voltage or load variations may
not produce exactly proportional speed changes in the motor,
so correction is approximate. An a.c. feedback signal is
controlled precisely by the motor speed so compensation is
more accurate.

Systems differ, but the basic principle consists of a
generator fixed to the rotor shaft. This can be a toroidal washer
divided into a number of segments, typically around thirty.
These disturb the field of a nearby permanent magnet as they
pass it, and a voltage is induced into a coil situated in the
magnetic field. Mutual interference between this and the rotor
driving field is avoided by displacing it by a 90° angle. The
frequency of the induced a.c. signal is the number of segments
multiplied by the rotational speed.

The output of the coil is applied across a frequency-
selective potential divider consisting of resistors and a
capacitor. As the reactance of the capacitor changes with
frequency, so does the voltage at the junction of the divider.
This is rectified and smoothed, then applied as a d.c. control
voltage to a current-control circuit supplying the motor. Speed
adjustment is obtained by making one of the potential divider
resistors variable.

Brushless motors   With d.c. motors, current is conveyed to the rotor by a commutator and brushes. Switching to alternate poles is abrupt, tending to produce a jerky rotation and noise is generated in the form of 'hash' that needs filtering to remove. A mechanical whine is often produced.

There are several types of brushless motor. Most use a permanent magnet rotor similar to the synchronous motor. In one type current applied to the stator windings is switched to alternate poles by the rotating rotor field. This is sensed by a pair of Hall-effect detectors, devices that change resistance in the presence of a magnetic field. The changing current through these is amplified and used to drive the stator coils. Like the commutator motor, speed is dependent on supply voltage, so any speed control circuit can be used with it.

Another type uses a set of three stator motor coils spaced at 120°, and a set of three secondary coils wound inside the first ones also at 120° but displaced 60° from the motor coils. There is also a single primary coil. The rotor is a permanent magnet with an aluminium disc at one end having a ferrite rod fixed across it. An oscillator feeds the primary coil which produces a field that is conveyed by the ferrite rod to whichever secondary coil is in range. The induced voltage is applied to an output stage that drives one of the motor coils. This turns the rotor 120° bringing the rod into the range of the next secondary. Thus a rotating field is obtained. Speed does not depend on the oscillator frequency. It is controlled by sampling the motor coil pulses, rectifying and smoothing them, then applying the resulting d.c. voltage to the output stages to control the motor-coil current. A pre-set is included for adjustment.

Speed adjustment   Timing a measured length of tape, then adjusting and re-timing until speed is correct is usually recommended. A less laborious method is to record a tone on two cassettes using a recorder of known speed accuracy; play one on that recorder, and the other on the machine to be adjusted. Adjust speed for the same pitch. As they approach, beats can be heard which slow to zero when the two tones are identical. For open reel machines, a tape stroboscope is the most effective speed check.

Hi-fi sound with video

Sound from video recorders is poor because although the video is laid down in high-speed helical tracks by heads on a rotating drum, sound is recorded at 1 inch per second in a linear track along the tape. Track width in both formats is 0.35 mm compared to 0.6 mm for audio cassettes.

FM recording   When a carrier wave is frequency modulated it deviates up and down from its standing frequency. The times per second it deviates is the modulation frequency and the extent of the deviation gives the modulation amplitude. A recorded f.m. signal is insensitive to amplitude variations, so is not subject to amplitude distortion or noise and the frequency response is not dependent on tape or head losses. Noise could be caused by wow and flutter, but 0.05% would produce only −66 dB noise. Bias is not required. The carrier frequency must be several times higher than the modulation frequency to achieve adequate deviation. While such a signal can be recorded at the high writing speed of the video heads it could not at the slow linear speed of the video tape sound track.

Beta hi-fi (American/Japanese)   The vision luminance (black and white) signal is frequency modulated between 3 MHz and 4 MHz, while the chrominance (colour) is amplitude modulated at below 1 MHz. The intervening gap is partly occupied by sidebands of both signals but only in areas of fine picture detail. Into this mostly free space, two f.m. carriers at 1 MHz and 2 MHz are interleaved, each modulated with one stereo channel. Sound is thus recorded and replayed by the video heads.

Beta hi-fi (European)   The above system is impractical for Europe because the PAL colour signal takes more bandwidth than the NTSC, and it has 625 picture lines instead of 525. So there is insufficient free space to fit extra signals. The European Beta system is similar to the VHS hi-fi.

VHS hi-fi   Two extra sound heads are used on the rotating drum, one each in front of a video head. These record two f.m. carriers modulated with a stereo signal at 1.4 MHz and 1.8 MHz, in a single track. Having wide head-gaps, they record deep into the tape coating. The video head follows behind, recording the video signals over the top of the sound track, thus partly erasing it. However, the 0.3 μm video head-gap field has but a shallow penetration into the 5 µm thick tape coating, so it erases only that part of the sound track near the surface. The deeper portions remain, resulting in a *depth multiplex* recording, having two layers.

To separate sound and vision signals on replay, the azimuth settings of the heads are displaced by 30°. The h.f. response of each head to signals of the opposite azimuth is well below the respective carrier frequencies, so there is a high rejection of the opposite signal. This is necessary in spite of the separation afforded by the sound and vision tuned circuits, because there is considerable overlapping of sidebands.

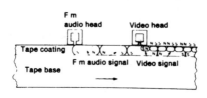

Figure 81. Depth multiplex recording.

Digital tape recording

Analogue/digital conversion   One method of conversion is by *successive approximation* whereby the analogue signal sample is given an approximate value corresponding to the most significant binary digit, then the remainder is approximated and assigned the next significant digit and so on to the least significant digit. Comparators are used to compare the signal voltage with that obtained from a ladder of resistors across the supply. Each comparator operates a switch that when on gives binary 1 and when off gives binary 0. A 'clock' circuit governs the speed of sampling which is up to 50,000 times a second (see CD section).

Digital processors   These devices permit a stereo digital audio recording to be made with a video recorder using either 14- or 16-bit quantisation. Any of the video recorder formats can be used. They consist essentially of a low-pass filter to prevent aliasing, an A/D converter and clock generator and a UHF modulator to feed the digital signal into the recorder's aerial socket. For replay, a video demodulator converts the television frequency back to digital, then a D/A converter restores it to analogue and the low-pass filter smooths out the quantising steps. Although a video recorder is used, the signal is *not* the same as the hi-fi video sound which is analogue modulated f.m.

Digital audio recorders   A number of these have appeared comprising a VHS videorecording mechanism with built-in digital processor. They use video cassettes but cannot record or play video.

Digital audio tape   A number of formats have contended for adoption as the standard for DAT recorders. There are two basic categories, the S-DAT which uses a stationary head with 20 tracks recorded in parallel simultaneously to achieve the necessary data rate, and the R-DAT, which uses rotary heads on a revolving drum in a similar manner to a video recorder. Each group has variations, though the cassettes are the same for each type. The proposed sampling rates were different from the CD record to prevent digital copying of records, but this has been revised in R-DAT 1, the one finally chosen. The reason is that recording companies can thus release a recording in both CD and DAT format from the same master tape. R-DAT 1 recorders have a 48 kHz sampling rate for recording, but a 44.1 kHz rate is automatically selected by pre-recorded tapes. Built-in encoding termed SCMS (Serial Copy Management System) prevents more than a first generation copy being made.

Digital tape formats

Format	Sample rate (kHz)	Number of bits	Transmission rate (Mb/s)	Recording density (Kb/in)	Writing speed (cm/s)	Tape speed (cm/s)	Channels
Analogue	–	–	–	50	4.76	4.76	2
S-DAT 1	48	16	2.4	64	4.76	4.76	2
S-DAT 2	44.1	16	2.206	64	4.37	4.37	2
S-DAT 3	32	16	1.6		3.17	3.17	2
S-DAT 4	32	12	1.2		2.38	2.38	2
S-DAT 5	32	12	1.2	(10 tracks)	4.76	4.76	2
R-DAT 1	48/44.1	16	2.46	61	313.3	0.815	2
R-DAT 2	48	16	2.46	61	313.3	0.9	2
R-DAT 3	32	12*	1.23		156.6	0.45	2
R-DAT 4	32	12*	2.46				4

*non-linear quantisation

The chart shows the formats with their main differences. The analogue compact cassette is included for comparison. Cassette sizes are: S-DAT (86 x 55.5 x 9.7 mm); R-DAT (73 x 54 x 10.5 mm); compact audio cassette (100.4 x 63.8 x 12 mm).

R-DAT 1   This is the standard format fur domestic digital tape recorders, although it is being challenged by the Philips DCC (Digital Compact Cassette). While the hardware is obviously different, signal processing is similar to that of the CD record. In the following description, reference back to the CD section where indicated, may clarify certain principles.

Track layout   Diagonal tracks are laid down on the tape by a pair of heads on opposite sides of a tilted rotating drum, similar to a video recorder. The drum rotates at 2,000 r.p.m., and the tape has a 90° wrap which is just a quarter of the drum perimeter. There are no gaps or guard bands between the tracks, and crosstalk is prevented by differences of azimuth between the heads, the angles being +20 and –20. Just as recorded h.f. information is not picked up by an off-azimuth analogue playback head, the track signal which is mostly h.f. is ignored by the adjacent head which is 40° off-azimuth to it. There are also two unused linear analogue tracks along each tape edge.

Each diagonal track is 23.5 mm long and 13.59 μm wide. It is divided into 196 blocks of data, each containing 288 bits, hence 56,448 bits in all. There are 8 bits to a *symbol*, which is the basic data unit, thus there are 36 symbols to each block. Quantisation is l6-bit so each signal sample consists of two 8-bit symbols. A pair of adjacent tracks are interleaved, and comprise one frame.

Reading along the track, each starts with an 11-block blank followed by 24 blocks of information and control sub-codes. Then comes 128 blocks of PCM (pulse code modulation) digital audio signal, and a further 22 blocks of information and control. Finally there is another 11-block blank.

Signal processing   The two channels of analogue audio signal are first pre-emphasised, then passed through a low-pass filter to remove all frequencies greater than half the sampling rate. Should these be present, the samples appear to the decoder to be values of a much lower frequency. because each successive sample sees a point of the next cycle but misses the intervening excursions. The effect is known as *aliasing*.

Next come the sample-and-hold circuits which sample the analogue signals from each channel and feed them in turn into the A-D converter.

Error correction   After conversion to a PCM digital signal, parity bits are added for error correction (see CD section). Each block also has added a symbol for sync, and symbols for identity codes and block address.

Then the signal is interleaved with that on the adjacent track so that the first track carries one odd left-channel sample alternated with one even right-channel sample; while its mate carries one odd right-channel sample alternated with one even left-channel sample. This permits interpolation of missing data if one track should be faulty. Interleaving is accomplished by the use of two 64 k RAM memories which can store one whole track each (56,448 bits),

These also enable time compression to be achieved. With a 90° tape wrap. the heads are free of the tape for 50% of the time and so not recording. The signal must be compressed into the remaining time along with the control and information codes. So signal is fed into the RAMs at an even rate but is read out in faster bursts between gaps.

After this the signal is applied to the cross interleaved Reed-Solomon encoder (see CD section). Two codes are employed. C1 and C2, the first effective for short error correction and the second for long bursts of loss or data corruption.

This encoder feeds data into a memory, and then reads it out in a different sequence. It is somewhat like a crossword in which all the 'across' entries are made, from which the 'down' answers are read off, or a matrix in which data is fed in horizontal lines and read out in vertical columns. Thus elements of individual symbols are spread over a wide area, and the effect of any damage may mean the loss of only one bit from a number of different symbols. The value of a single missing bit can be accurately averaged from the two adjacent ones.

The total effect of all these correction systems is that over 2.5 mm of damaged track length can be handled with little audible effect. This length contains over 6,000 bits. Even more can be dealt with by interpolation from the adjacent track.

8–10 Modulation   On leaving the encoder, the digital signal is passed to the 8–10 modulator which is similar in effect to the 8–14 modulation of the CD (see CD section). It is not a modulator in the generally accepted sense of imposing one signal upon another. but rather an exchanger. It changes the 8-bit words into 10-bit. using an electronic 'dictionary'. The new words avoid long strings of 1s or 0s which otherwise occur in some binary numbers. that could produce d.c. signals difficult for the circuits to handle.

Sub-codes There are two distinct types of sub-code. the first. termed the PCM-ID, identifies technical data about the recording and is part of each PCM block. The second, called ID, contains user information and control data. ID sub-codes are in separate blocks arranged at the start and finish of the PCM string of blocks in each track. They can be changed, whereas the PCM-IDs being part of the PCM blocks cannot.

PCM-ID The first seven are 2-bit codes defining 1/ use (audio, data storage etc.); 2/ pre-emphasis on or off; 3/ sampling frequency; 4/ number of channels, 2 or 4; 5/ quantisation; 6/ track pitch, (normal or wide for some pre-recorded tapes); 7/ copy prohibit on or off.

The eighth is termed a pack and consists of 32, 2-bit codes. These are at present not used but reserved for future applications.

The ninth and tenth codes are optional; that is. they can be used or not as the recorder manufacturer sees fit, and no doubt this depends on the price of the model. All the previous ones are mandatory.

Code nine which has four, 4-bit numbers is the SC (search code), and carries the track number, and elapsed time in hours minutes and seconds. Code ten also having four, 4-bit numbers is the AC (auxiliary code) used mostly for pre-recorded tapes; it conveys programme information.

All PCM-ID codes are carried by 8 bits included in each PCM block, with 128 blocks this gives 1,024 PCM-ID bits per track. Each code is thus repeated a number of times along the track.

ID There are eight ID sub-code blocks near the start of each track after the eleven-block blank and two phase lock blocks, and eight at the end before the final sync block and eleven blanks, so making sixteen blocks of ID per track. These contain control and data information, but unlike those appearing in the PCM blocks. they can be changed by the user.

One control ID is the Start Code which marks the commencement of a recording. It is repeated over the first nine seconds of any recording, and enables the section to be located in the fast search mode. As the 9 second passage is scanned in $^1/_{20}$ of a second and tracking errors arise due to the changed head-to-tape speed it can be seen why the Start Code is so repeated.

Another ID is the Skip Code which instructs the machine to skip to the next Start Code. This can be pre-set by the user if a particular section is not required. A further one identifies a pre-recorded tape.

Programme numbers are contained in the PNO-ID codes which allow for editing or re-numbering the sections. Consecutive numbering is already contained in the PCM-ID blocks, but these can be ignored in favour of the PNO-ID code set by the user so that any pre-selected collection of items can be played.

Automatic track following There are five blocks of ATF signals before and another five after the PCM blocks in each track. Each are sandwiched between two spacers (called IBG, interblock gap) consisting of 3 blocks of a 1.568 kHz signal.

The lateral speed of the tape is such that the recording heads erase a strip of a third the width of the previous track. When playing back, this makes the head gaps 1.5 times wider than the recorded tracks, so they overlap the two adjacent ones. The high frequencies of the digital PCM signal are not picked up as crosstalk because of the azimuth difference, but the lower ATF pilot tone is unaffected by it.

There is an ATF switch-on signal in each ATF section, a different one for alternate tracks, These are 522.67 kHz and 784 kHz. Their function is to activate the sampling and correction circuitry.

The pilot tone which appears in all tracks is 130.67 kHz. alternating with gaps containing the 1.568 kHz spacer frequency. When the ATF is switched on, the head starts reading the pilot tones from adjacent tracks. The blocks are so aligned that when a pilot tone is present in the track to the left, there are spacers in the right-hand track and the track being followed. Next, the pilot tone appears in the track to the right with spacers in the left-hand and followed tracks.

Thus the pilot tone appears on alternate sides of the followed track and the levels received are compared. If not equal, a signal is sent to the capstan servo to change the tape speed and thereby the tracking angle.

Serial Copy Management System   (SCMS) A single copy can be made from a digital original, but this is automatically given an inhibit-copy value in the PCM-ID copy-prohibit code. Recorders will thus not make second-generation copies although any number can be made from the original.

DAT modes   There are six DAT modes of which three are optional. The three compulsory ones must appear in all models.

● **Mode I** This is the standard one that uses 48 kHz sampling frequency and 16-bit linear quantisation.

● **Mode 2** An optional mode which has a 32 kHz sampling frequency but otherwise the same as mode 1. It is designed to be able to record satellite digital stereo broadcasts.

● **Mode 3** A half-speed optional mode for doubling recording time. Sampling is at 32 kHz, and quantisation is 12-bit non-linear (see CD section). Tape speed is 0.4075 cm/sec, and playing time up to 4 hours.

● **Mode 4** An optional four-channel mode at the same quantisation and sampling frequency as mode 3. but with normal running times.

● **Mode 5** Intended for playing pre-recorded tapes. it has a 44.1 kHz sampling frequency and 16-bit quantisation. The recording industry was initially opposed to DAT recorders with the same sampling and quantisation as CD records, as this would allow copying. They then found that having different systems would mean re-mastering their own recordings if released on both CD and DAT.

Unwilling to do this or forego sales of pre-recorded tapes, they withdrew objections but at first wanted the mode to be playback only. The SCMS copy-prohibit code has nullified the reason for this so the mode has full record and playback capabilities.

● **Mode 6** An alternative pre-recorded tape format using a 20.41 μm wide track and an increased 12.25 mm/s tape speed. It allows the use of barium ferrite tape instead of metal, but it has a maximum playing time of only 80 minutes. The mode is mandatory.

Cassette The size is 73 x 54 x 10.5 mm: it has a sprung lid and a sliding shutter to cover the hub holes giving total protection when out of the machine. There are no guides or rollers. the tape rests across a pair of round moulded extrusions, from which it is pulled clear when loaded.

A prism is mounted at each end of the operating aperture. which reflects light from an LED in the machine through the transparent tape leaders and so actuates the end-of-tape stop.

There are 5 ID holes at the back of the cassette. Three identify the type of tape: all open signifies standard metal tape; the centre one open, thin LP metal tape; third one open, standard BaFe tape; second and third open, thin BaFe tape.

Hole four open signifies a pre-recorded tape, while hole five open prevents recording in the same way as in the compact audio cassette.

Tape The standard tape is 3.81 mm, the same as the compact cassette, with metal formulation. Base thickness is 9.5 μm; coating. 13 μm; and back-coating 0.5 μm; a total of 13 μm. Coercivity is l,500 oersteds. It gives two hours' playing time.

An alternative is barium ferrite (BaFe) having a coercivity of 7000 oersteds. This lends itself to easier copying for pre-recorded tape manufacture. by heat contact transfer from a master.

Philips digital compact cassette

A late contender in the digital tape stakes, it is claimed not to
rival DAT because it is a mass-appeal format, being to DAT
what the compact cassette was to open-reel systems. It is
envisaged to become the popular digital tape format, leaving
DAT the choice of professionals and 'high end' users.

However, it has many attractive features and it would be
surprising if it did not become the accepted digital recording
medium just as the audio compact cassette has, with DAT
fading eventually into obscurity.

Complexity   The DCC system uses a single stationary head
and thereby eliminates the complication of a two-head rotating
drum, with its sync and servo systems, and also the head wear
resulting from fast tape contact. It simplifies loading, as the
tape is not pulled out from the cassette but is played in the
same manner as an audio compact cassette.

The DCC linear tracks are 185 μm wide with 10 μm gaps,
compared with 13.59 μm of the DAT system, with no gaps.
Successful playback requires a width of only 70 μm, so there is
a considerable tolerance in the playback head positioning. This
eliminates the need for track-following signals and control
circuitry, and ensures that tapes recorded on one machine will
be compatible with. and playable on others.

The tape speed of 4.76 cm/sec enables sub-codes to be
recorded on a separate linear track instead of being mixed with
the digital audio as with DAT. The DAT tape speed of
0.815 cm/sec is too slow to permit this.

Interleaving of different samples is carried out with DAT
and CD in order to spread the effect of local tape or disc
defects. By using a number of parallel tracks each carrying part
of the signal, DCC achieves a similar protective effect, thus
obviating the need of electronic interleaving with its
requirement of large RAM memory stores.

The encoder is more complex than DAT as will be seen, but
it is contained within two chips, so the external encoding
circuit complexity is little different from DAT.

Compatibility   A remarkable feature is that DCC is fully
compatible with ordinary audio compact cassettes. The DCC
cassettes are the same size and shape, and the tape is the same
width. The tape speed is also the same. Thus an audio analogue
cassette can be played in a DCC machine, the head being
designed to accommodate DCC and audio stereo.

Furthermore. the shortest recorded wavelength of 0.99 μm
is recordable on chrome tape so in theory, a conventional

chrome cassette could be used to record DCC. Very likely, cobalt-doped ferric tapes will prove an acceptable alternative. DAT has a shortest wavelength of 0.67 μm which requires expensive metal tape.

DCC cassettes though basically similar, have hub holes on only one side because the system uses auto-reverse. Programme details can thus be printed over the whole of the opposite side.

They also have a cover over the tape aperture which slides away when the cassette is inserted. This protects the tape when in a hostile environment.

System requirements   The obvious problem is how to squeeze a high sampling-rate, high-quantisation digital signal into a compact cassette format using a single stationary head, and a 4.7 cm/sec tape speed, yet achieve a 90 minute playing time from a C90, the same as for analogue.

The DCC head   The DCC sector, which is the upper half of the head where the two analogue stereo tracks would normally be, has eight, 185 μm head gaps. These are a product of thin-film construction techniques. When the end of the tape is reached, the head rotates through $180°$ so that the eight-track sector now records the bottom half of the tape.

The lower half of the head is the analogue sector, and this contains two conventional stereo gaps. For analogue playback, the head is orientated so that these are at the top for the start, then rotated to the bottom for the reverse run.

PASC encoding   Reference to Figure 3 on page 3 reveals that the threshold of hearing indicated by the bottom plot, is greatly dependent on frequency. All the blank areas below the plots are inaudible. Any sounds in this region thus do not need to be encoded.

A quiet sound is masked by a loud one of similar frequency, so these are redundant and not required.

There are large gaps of silence at various changing frequencies during musical performances, and the whole sound spectrum is never reproduced at the same time except with white or pink noise. These gaps also need not be encoded.

These three factors reveal that a lot of wasted encoding takes place with conventional digital systems. DCC encoding digitally splits the spectrum into 32 third-octave segments and monitors the activity in each. Any gaps, masked frequencies or those below hearing threshold are not encoded, and unused assigned bits for that segment are re-assigned elsewhere where they are needed.

The result is PASC (Precision Adaptive Sub-band Coding), which gives an equivalent of 18-bit quantisation at only a quarter of the bit requirement for normal 16-bit digital encoding. The 32 segments are mixed down to eight channels for recording the eight tracks.

Error correction, modulation and SCMS   These are similar to the DAT systems. C1 and C2 Reed–Solomon correction is used with 8–10 modulation. The SCMS copy-prohibit code is also used to prevent digital copying of more than one generation (see DAT section).

Modes   There are three sampling rates available, 48 kHz. 44.1 kHz, and 32 kHz. Thus pre-recorded cassettes can be produced from the same masters as CD and DAT records. Quantisation is variable according to segment signal level, but can be up to 18 bits, thereby giving greater potential definition than either CD or DAT.

Frequency range at 48 kHz sampling rate is 5–22,000 Hz, while the dynamic range is 105 dB. Total harmonic distortion including noise is –92 dB.

Recording tape

Seemingly simple, recording tape is a highly complex product. There are many factors which govern its performance; the basic materials used being a major one.

Base   Many materials have been used for the tape base including acetate, p.v.c., and even paper. Polyester (polyethylene glycol terephthalate) is now universally used being superior to all others used so far and available at moderate cost; it will stretch slightly under stress. Other materials such as polyimide are better, but are not at present obtainable in quantity and at comparative cost.

The thickness of the base film varies from typically 35 μm for open reel tape, 12 μm for a C60, 7.5 μm for a C90 and 6 μm for a C120 cassette. Below 12 μm, the film needs to be *tensilised* to withstand the stress. Normal film is stretched equally in both directions when extruded and is said to be *balanced*. Tensilised film is unequally stretched so as to increase longitudinal tensile strength at the expense of the transverse.

Roughness of the film is imaged through the coating to the tape surface; it can modulate the signal causing noise, drop-out and can increase head wear. Smoothness is thus essential. While optical smoothness is possible, it could exclude air between layers causing distortion of the film during manufacture and spooling problems and tape tangles in use. A *controlled roughness* is therefore introduced by suitable additives. Alternatively, a rough and smooth roller can be combined for the extrusion pair so that the back is given a rough surface and the front is made smooth, A backing coating, which in addition can be made anti-static, is another method.

Ferric coating   Not all iron compounds are magnetic, only those in which the various directions of their electron orbits do not completely cancel, that is those having a cubic atomic lattice structure. Magnetite, Fe_3O_4, has such a structure consisting of 24 iron atoms and 32 oxygen atoms. But it is chemically unstable being halfway between the simple low oxide of iron FeO, and the high oxide Fe_2O_3. It thus makes an unreliable tape coating. However, alpha Fe_2O_3 having 5 atoms cannot form a cubic lattice, but if prepared by further oxidising magnetite it can be made so. Then it forms a lattice of 32 oxygen atoms and an average of $10^2/_3$ iron atoms. The result is a superlattice, some having 10 and others 11 iron atoms in a systematic, hence stable structure. This form is termed *gamma Fe_2O_3* (Fe_2O_3 is ferric oxide; Fe_3O_4, ferrous oxide; FeO, iron oxide).

Dopants   Gamma Fe_2O_3 still has vacancies in its lattice so its magnetic properties can be considerably enhanced by adding small amounts of a dopant such as cobalt to occupy them. In particular, coercivity can be increased from around 300 oersteds to nearly 1000 according to the amount added. Early cobalt-doped tapes suffered from unstable magnetism and print-through due to uneven distribution of the dopant. Modern techniques employ *epitaxial* oxides in which the dopant is diffused into the surface areas of the gamma oxide particles afterwards.

Particle shape   The predominant field in the head-gap is longitudinal, so thin lengthwise particles in the tape coating give it a higher coercivity than round or random shapes. The magnetite from which the gamma Fe_2O_3 is derived is obtained by dehydrating hydrated iron oxide FeOOH. This in turn is made by an aeration process whereby seeds of the compound are grown. These are needle-shaped or *acicular* which is almost ideal. Length, which is controlled by seeding time. is from 0.1 to 0.5 μm.

Hydrothermal process   This is an alternative production method which uses a high pressure of several hundred atmospheres and heat on alpha Fe_2O_3 instead of seeding hydrated iron oxide and converting to and from magnetite. This produces particles which are ellipsoid-shaped like a lengthened rugby ball, which are even better than the needle-shape.

Chromium dioxide   The hydrothermal process was originally used to produce a magnetic form of CrO_2 having a tetragonal lattice, then latterly applied to ferric oxide. CrO_2 is magnetically little superior to gamma ferric oxide, but the hydrothermal process gave more uniform particles of better shape, hence a superior h.f. response. The advantage is offset now ferric oxide is being formed by the same process, especially when doped with cobalt. It needs 6 grams of CrO_2 to produce a C90 cassette.

Pure metal   Pure iron, having superior magnetic properties to its oxides, seems the obvious and perhaps ultimate choice of tape material. The major problem is that the very small particles needed oxidise immediately they are exposed to air, sometimes with explosive force! This has been overcome by finely covering each particle with a protective coating. Other ferro-magnetic metals such as cobalt, nickel, chromium or their alloys could also be used.

Binders   Binders consist of one or more polymeric resins chosen for their strong adhesion to the base film, with the qualities of toughness and flexibility. A compromise between these is obtained by blending suitable proportions of brittle and elastic resins. These are combined with various additives to improve flexibility, reduce static and to lubricate it. A solvent is also needed to combine all into a liquid dispersion. The binder and additives make up about 60% of the coating, the magnetic gamma ferric oxide powder only some 40%.

Wetting agent   The magnetic powder is inorganic in nature whereas the resins are organic, so they are incompatible. A wetting agent is therefore required as an interface. It must also expel all air between particles and permit the ingress of solvent. Such agents are partly organic with long chains of carbon atoms that combine readily at one end with the organic binders, while the other end has an affinity for inorganic ferric oxide.

Every oxide particle is completely coated with the agent to a thickness of one molecule, having its organic end outward.

This requires a large quantity of agent because the particles are so finely divided that they have a large surface area. A pound weight of particles has a surface area of about a quarter of an acre.

Thin-film coatings Metals can be deposited directly onto plastic film without the use of binders so achieving 100% concentration. Three methods are possible. *Electroplating* with a primer coating and with small amounts of phosphorous to provide a grain structure so that magnetic zones can be formed; *evaporation* from boiling metal in a high vacuum to reduce the temperature; and *ionic sputtering* by bombardment of a metal cathode by positive ions of an inert gas.

All have their snags. Electroplating is very slow; the high evaporation temperature of some 2000°C can melt the tape, so it needs to be done very fast with several passes to build up the layer; sputtering is slow, energy-inefficient and wastes much material.

Films so produced are 0.1–0.3 μm thick which is ideal for recording very short wavelengths that self-demagnetise in thick coatings, but give low output and high noise at audio frequencies. Ideal applications are thus video and digital audio. Video evaporated-film tapes are in production.

Tape manufacture

Milling The ferric oxide powder, being magnetic, clumps together and so is received as an aggregate. It must be broken apart into its individual particles, without breaking the particles themselves. These are already of the correct size and shape and, if made smaller, will magnetise too easily, making the tape liable to print-through. The operation was at one time performed by a *ball mill*, a slowly rotating cylinder containing steel balls into which a batch of aggregate was placed. Results were uncertain, some particles being sheared too much, and others not at all.

Bead milling A large number of very small glass beads up to 1 mm diameter are whirled turbulently in a chamber through which the aggregate passes, so offering a large number of contact situations. The process is continuous, the aggregate passing in turn to other mills having successively smaller beads. Additives, binder and solvent are added along the way; filters being interposed to remove the beads. Solvent quantity is carefully controlled: too much causes coating shrinkage after drying and too thin a dispersion; too little produces filtering problems.

Coating   The polyester backing film is presented to the coating head in rolls 5 ft (1.5 m) wide and some 10,000 ft (3000 m) long. There are two principal methods of coating: *reverse roll* and *gravure*. With reverse roll, solution is fed to the applicator roller while a contra-rotating metering roller, spaced precisely the coating thickness away, removes the excess which is scraped off it by a blade held against its surface. The film passes between the applicator roller, which rotates in the opposite direction to the film flow, and a pinch roller. It thereby receives the exact amount of coating material.

With gravure coating, the applicator roller has spiral grooves etched in it that have the exact capacity of the needed solution. When all surface material is removed, that remaining in the grooves is deposited on the film which is pressed against its surface by a slightly compressible pressure roller. The groove pattern is smoothed out mechanically or magnetically. The coating thickness cannot be changed with gravure coating, but it can with reverse roll by adjusting the roller spacing. Accuracy of roller concentricity is essential with reverse roll, as any eccentricity produces cyclic variation of coating thickness. Coating thicknesses are: open reel, 10 μm; C60, 6 μm; C90, 5.0 μm; C120, 3.5 μm.

Orientation   The needle-shaped particles must all be aligned along the tape if they are to be effective. This is done by passing the film through a magnetic field before the coating dries. The orientating field must not be so strong as to cause a violent particle movement and thus roughen the coating, but must be sufficient to ensure all particles are in the correct plane. Failure could result in the *velour effect*, whereby tape performance differs when recorded in one direction from that when recorded in the other.

Drying   Drying is carried out in a heated drying tunnel from which air and the solvent vapour is removed. Temperature and air flow is carefully controlled so that evaporation of solvent from the coating is gradual and so unlikely to create voids in the surface. To avoid contact with anything that could mar the surface the film is floated down the tunnel on jets of air rather than on rollers. Jets above and below keep it in a central position throughout its passage. As the solvent/air mixture is explosive, safety precautions must be taken in its evacuation.

Calendering Surface finishing is achieved by passing the coated film between sets of calendering rollers. One of each pair is hard and is heated, the other is slightly compressible. Pressures of some 2300 kg per cm width are applied at temperatures of 80–120°C depending on the exact composition of the binder.

At least three purposes are served by this action. One is compaction of the coating; the magnetic material is made as dense as is possible to get it, and microscopic cavities are crushed out. The second is polishing; the surface of the compressible roller de-accelerates and accelerates as it goes through the area of contact, so giving a rubbing action as it takes the film past the hard roller. Thirdly, lubricants and other additives are squeezed up to the surface where they are needed. Finally, curing of the binder and crosslinking of the resins is increased.

Curing Curing and crosslinking of the binder needs to be at an advanced stage before the roll of film is subjected to slitting. It proceeds exponentially with time so is never fully completed, but is sufficiently progressed after a few days to continue. The rolls are therefore stored in a warm environment for several days or even weeks after calendering. Instant curing by electron beam bombardment is an alternative method, but it requires different types of resin and has other practical difficulties.

Slitting The roll is slit into individual tapes by two sets of rotary cutters. These enmesh from both sides of the film, penetrating it slightly, each cutter having a slight side pressure against its mate in the other set. The operation of each pair is like that of a pair of scissors but with continuous rotary action. Laser cutting is another method, but choosing the right energy level to penetrate the transparent film and opaque coating requires judicious compromise. The take-up spools are driven by a slipping clutch to maintain correct winding tension.

Cassette assembly Measured lengths are wound off a large pancake of tape and the leaders added by an operative who then assembles the hubs, rollers, pressure pad, shield and liners into the cassette shell and screws on the upper half. Special tools and jigs facilitate the job, but the human eye and hand still play a major part.

Musicassette manufacture

Mother tape   A mother tape is dubbed down at $7^1/_2$ i/s from the studio master and encoded with Dolby B noise reduction. A coded signal is also recorded at its start and end. It is played at 32 times the recording speed as a continuous loop, and fed to a bank of slave machines running at 32 times normal playing speed of $1^7/_8$ i/s, or 60 i/s. Each slave records several consecutive performances on a single long tape run.

C zero cassettes   Cassette cases known as C zeros have a single length of leader fixed to both hubs. This is cut by the operator who splices in the start of the slave tape. The tape is wound at speed into the cassette over a head that senses the end-of-performance coded signal. Here, the tape is stopped, and the precise finish located. Next the tape is cut and the end spliced to the remaining portion of leader.

After winding in the leader, the cassette is ready for labelling and boxing. Samples are tested, and members of the same batch are kept together so that others can be readily investigated should a fault be found in a sample.

Tape parameters

Bias   A plus or minus dB value is sometimes given, this being relative to a DIN standard for the group, or a specified reference tape. Tapes of similar grade and bias figures will give similar results on fixed bias machines. A tape with a lower bias figure than that normally used will be overbiased and so have lower distortion but poorer h.f. response, whereas a higher figure will give better h.f. response with higher distortion.

Coercivity   The reverse magnetic force necessary to reduce the remanent field of a tape magnetised to saturation, to zero. High coercivity tape is less susceptible to self-demagnetisation or demagnetisation from external fields, but needs high recording and erase currents. Unit is the oersted or kAmp/metre (1 kA/m = 12.5 oersted).

Distortion   Most of the distortion generated in the process of tape recording is third harmonic so the quoted figure is for that. The figure should stipulate the recording level (usually 0 dB, or –3 or –4 dB); it can be assumed it is for the optimum bias level. Sometimes a graph displays the distortion for different levels of bias.

Dynamic range   The difference between the largest and smallest recordable levels. The largest is ultimately limited by tape saturation, but prior to this the curvature of the hysteresis loop produces compression of the signal and distortion. The maximum output level for a given distortion thus sets the top limit. Noise level establishes the lowest. To be unobtrusive, noise should be at least 10 dB lower than the quietest recorded sounds, so in practice the usable dynamic range is some 10 dB less than the calculated amount.

Frequency response   H.f. response is poor at high levels owing to various h.f. losses. Response is therefore usually given at –20 dB.

MOL   Maximum output level, sometimes called MML, maximum modulation level. It is the maximum level for a given distortion, usually 3%, but sometimes 5%. High frequency MOL is often given for saturation. The standard MOL is specified for 315 or 333 Hz, while the h.f. MOL is for 10 kHz. Both are highly dependent on bias, the standard increases with bias increase, whereas the h.f. decreases.

Print-through   The transference of a recording to adjacent tape layers producing echo or pre-echo, when the tape has been wound without re-spooling for a period; specified as the ratio in dB between the original recording and that transferred. Smaller particles are more susceptible, so it is less of a problem with lower grade tapes.

Reference level   The 0 dB level to which MOL and sensitivity are related, corresponds to 250 nanowebers/metre recording flux. This should give a 0 dB reading on the VU meter. The Dolby reference is 200 nW/m, or just under –2 dB. A level of –3 dB is 177 nWb/m.

Remanence   This specifies the amount of magnetism left on the tape after the magnetising force has been removed. It thus directly affects the sensitivity. Unit is the gauss or millitesla (1 mT = 10 gauss).

Sensitivity   The output at various frequencies obtained for a recording level of –20 dB (25 nWb/m). Usual frequencies used are: 315 Hz, 3150 Hz, 10 kHz and 16 kHz.

Tape groups

Blank tapes are divided into four groups depending on the type of coating and its recording characteristics. These vary considerably between brands; those given here are a rough guide.

Group 1   (ferric types with 120 µs equalisation) This is the most common type being also the variety used in the manufacture of musicassettes. There are three basic grades. Lowest is the type variously designated 'standard', 'dynamic', 'ferric' and 'low-noise' They have the lowest remanence and signal/noise ratio, but improvements in every grade has brought these up to the standard of high-grade tapes of a few years ago. The best are capable of quite good results. Bias level requirements are usually 1 or 2 dB lower than the higher grades in this group. Suitable for portable recorders and average music centres. Typical coercivity, 380 oersteds for all in this group: remanence, 140 mT for this grade.

Next grade are those using smaller more densely packed particles often termed *microferric*. This improves overall sensitivity by about 1 dB, rising to over 2 dB at the h.f. end: MOL is also some 2 dB higher. Noise is slightly improved, but print-through is usually worsened by 3–6 dB. Remanence is around 160 mT.

Highest grade are often termed *high energy* types, and are sometimes lightly doped with cobalt. Sensitivity is some 2 dB overall rising to 3.0 dB at h.f. greater than the standard tape, with MOL around 4 dB higher. Noise is similar to the second grade, but the higher MOL gives a corresponding improved signal/noise ratio. Remanence about 175 mT.

Group 2   (chrome and cobalt ferric types with 70 µs equalisation) The good h.f. response of CrO_2 requires less h.f. equalising boost on playback, so boost is applied from 2.2 kHz instead of from 1.2 kHz as with group 1 tape. Thus noise is also reduced, but some 3 dB extra recording level is needed to compensate for the lower playback boost. With early chrome tape low frequency response was poor and the tape was abrasive, causing excessive head wear.

These problems have been largely overcome, but chrome has fallen from grace in favour of cobalt-doped ferric tapes having similar characteristics and intended for use on 'chrome' switch positions, As with group 1, different grades exist with similar characteristics, though with slightly increased h.f. response. The main differences are 3–5 dB less noise (owing to different equalisation) and higher coercivity at around

650 oersteds. H.f. MOL is often some 2 dB lower than comparable group 1 tapes, which partly offsets the lower noise figure.

Group 3   (ferrochrome) Owing to the poor low-frequency response of chrome, a dual layer tape of chrome on top of ferric was developed; the chrome to carry the h.f. and the ferric to take the l.f. It is now virtually obsolete.

Group 4   (pure metal tape with 70 µs equalisation) This has a typical coercivity of 1100 oersteds, and a remanence of 330 mT. The h.f. MOL. is some 4–10 dB higher than the highest non-metal tape and there is a better balance between l.f. and h.f. MOL. Higher bias (3–6 dB higher than chrome) and erase currents are required. Some types are designed to work as group 2 tapes having lower remanence and coercivity than standard metal, and an h.f. MOL slightly up on non-metal tape. Some group 4 tapes use metal alloy formed into 0.03 µm balls strung in chains.

Open-reel recorders

Once the standard domestic tape machine, open-reel recorders are much less in evidence now. They are still used by enthusiasts for some of the special features not otherwise obtainable.

Tracks   Either *half-track* (one track each way), or *quarter track* (two interleaved tracks each way; 1 and 3 forward, 2 and 4 reverse). Stereo machines are usually quarter track but some are half-track.

Speeds   Single speed machines run at $3^3/_4$ i/s, but three-speed models used $1^7/_8$, $3^3/_4$ and $7^1/_2$ i/s. The standard 1200 ft tape runs for 1 hour at $3^3/_4$ i/s.

Editing   Ease of editing is a feature of open-reel machines. The precise editing point can be located on the tape by manually turning the spools with the machine in the pause mode, then the point on the playback head can be marked with a chinagraph pencil. An editing block secures the tape while a straight or diagonal cut is made along a groove. The two ends are joined by special jointing tape (ordinary adhesive tape should *not* be used).

Duoplay/sound with sound   A recording made on track 1 of a quarter track machine is played back via headphones while a further recording is made on track 3. Both can then be replayed together by means of a parallel track switch. Thus one person can record a duet.

Sound on sound   A recording on track 1 can be played and monitored by headphones while another is recorded. *Both* are recorded together on track 3. Track 3 can then be played and monitored while a new recording is made, the combined result appearing on track 1, so erasing the original. The process can continue indefinitely, each new recording being added to what has gone before. If the latest part is faulty it can be re-done as the recordings up to that point are still intact on the previous track. The practical limit is set by the addition of noise and distortion at each re-recording which becomes noticeable after about four transfers. Bass and percussion parts are best recorded first as these show distortion less than treble parts.

The compact cassette

Originally designed by Philips, the compact cassette has developed into the standard high-fidelity domestic recording medium.

Tracks   Unlike open-reel stereo, each stereo pair is not interleaved, but adjacent. The outer tracks are the LH channel and the inners the RH. Each is 0.6 mm with a 0.3 mm gap between each channel and a 0.6 mm gap in the centre between the inners of each pair.

Speed/timing   Tape speed is $1^7/_8$ i/s (47.625 mm/s). Tape length of a C90 cassette is 135 m, giving a playing time of 47.25 minutes per side. A C60 is 90 m in length with a playing time of 31.5 minutes per side.

Protection/identification apertures   A removable tab at each end of the back of the cassette can be broken off to permit a feeler to engage and so block the record key. Thus valued recordings are protected against accidental erasure. The correct tab for a particular side is the one nearest the full spool at the start of play. Adjacent to this is an aperture on group 2 cassettes to give automatic bias and equalising change with some decks. Further in is a similar aperture for metal tape.

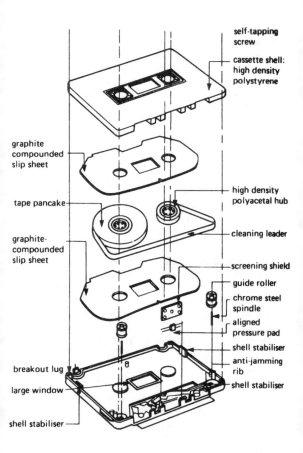

Figure 82. Exploded view of a compact cassette.

High-quality radio

Receiver principles

Amplitude modulation   When a high frequency is modulated by an audio signal, it is termed a *carrier* and the audio is the *modulation frequency*. If the shape of the carrier varies with the modulation it is said to be *amplitude modulated*. Other frequencies are produced which correspond to the carrier plus and minus the modulation frequency. These are known as *sidebands*. The total range of frequencies from the highest to the lowest sidebands describes the *bandwidth* of the signal.

SSB   In the radio-frequency spectrum, each transmission thus represents not a single frequency but a band which depends on the extent of the modulation frequencies. To economise in band space and transmitter power, one sideband is sometimes suppressed as the modulation can be adequately conveyed by the other. This is termed *single-sideband* (SSB) transmission. This requires special receiving circuits and is usually confined to communication applications. Another method is to suppress part of one sideband and is termed *vestigial sideband* transmission.

Receiver bandwidth   The receiver-tuned circuits must have a bandwidth equal to that of the transmitted signal, otherwise the higher audio frequencies will be lost. If the bandwidth is excessive, sidebands from adjacent transmissions will be received. The degree to which a receiver rejects adjacent transmissions is termed its *selectivity*. However, as high selectivity means reduced bandwidth and audio h.f. loss, some a.m. receivers have *user-variable selectivity*.

TRF   Earliest receivers were *tuned radio frequency* (TRF) devices in which a number of tuned circuits separated by amplifying stages were all tuned to the transmission frequency. The greater the number of tuned circuits, the higher the selectivity. Tracking was a problem as all the circuits were tuned together by a ganged variable capacitor, as was instability, owing to stray coupling at the same frequency.

Superheterodyne   Now universally employed, the superhet usually has one stage of amplification and tuning at the transmission frequency called the *radio frequency* (r.f.) stage, followed by a *frequency changer*. This consists of an oscillator that is tuned by the variable tuning control to the transmission frequency plus a fixed frequency. When mixed with the output of the r.f. stage, frequencies are produced equal to the sum and difference of the oscillator and r.f. frequencies. The difference is called the *intermediate frequency* (i.f.), which remains the same whatever the frequency of the received transmission. Intermediate frequencies for a.m. radios are usually from 455 to 470 kHz, but others are also used. Subsequent amplifying stages thus need only fixed tuning circuits. The desired bandwidth can be obtained by *stagger tuning*, whereby the i.f. circuits are tuned to slightly different frequencies.

Demodulation/detection   The final carrier whether at r.f. or i.f. is rectified, so suppressing the negative half-waves. A small capacitor bypasses the carrier frequency in the positive half-waves leaving the audio signal for further amplification.

Shortcomings   Although capable of high quality sound, the crowded nature of the medium waveband has led to severe bandwidth restriction for each transmitter to 9 kHz, which gives a modulation frequency of 4–5 kHz maximum. Sidebands of unauthorised transmitters cause jamming; a.m. also is vulnerable to electrical interference.

Frequency modulation   The carrier is of fixed amplitude but the frequency varies above and below that of the unmodulated carrier in sympathy with the audio modulation. The amount the frequency varies, termed the *deviation*, conveys the amplitude of the modulation; thus a soft audio signal produces a small deviation, whereas a loud signal generates a large one. The standard employed by the BBC is a deviation of ±75 kHz which corresponds to 100% modulation while ±22.5 kHz is 30%.

Frequency of the modulation is conveyed by the deviation rate, i.e. a 1 kHz tone will cause the carrier frequency to deviate 1000 times a second in each direction.

Intermediate frequency   Standard nominal i.f. for f.m. is 10.7 MHz.

Receiver bandwidth   A maximum deviation of ±75 kHz requires a minimum i.f. bandwidth of 150 kHz, but to allow for mistuning or oscillator drift, the bandwidth is usually 230 kHz.

Receiver tuner   Unlike a.m. circuits in which the r.f. stage is often omitted, one is always used for f.m. It provides gain ahead of the frequency changer and thus reduces noise; it also provides a buffer to isolate the aerial from the local oscillator which could otherwise radiate interference to other receivers. An acceptor circuit (series coil and capacitor) is shunted across the input of the frequency changer (also termed the *mixer*) to bypass any 10.7 MHz signal that may be picked up by the aerial and passed through the r.f. stage to the mixer and i.f. stages.

Transistors in the r.f. and mixer stages are usually operated in the common base (or gate for f.e.t.s) mode although common emitter circuits are also used. Tuning is generally by a twin gang variable capacitor, or varicap diodes, but variable permeability coils with sliding cores are sometimes employed. Capacitors across the oscillator tuning circuit frequently are split between positive and negative temperature coefficient types, so that capacitance changes with temperature are balanced out.

I.f. stages   The i.f. stages need to be of higher gain than for a.m. because of the wider bandwidth requirements which reduce the stage gain. More discrete stages are used or a high-gain integrated circuit. The tuned circuits are provided by either i.f. transformers or ceramic filters, or both. Transformers are generally used in the output of the mixer stage, while filters appear between successive i.f. stages.

Ceramic filters   The filters mechanically resonate at the i.f. and so pass on only a limited band around that frequency. The frequency is more stable than the i.f. transformer and no adjustment is required. Bandwidth is typically 300 kHz for each filter and there is an insertion loss of about 6 dB. The centre frequency is not always exactly 10.7 MHz. Receiver performance is not affected by minor variation of the i.f. but all filters used must have the same centre frequency. They have a general tolerance of ±30 kHz and are colour coded as follows:

Black	10.64 MHz
Blue	10.67 MHz
Red	10.7 MHz
Orange	10.73 MHz
White	10.76 MHz

Pre-emphasis   High frequencies are boosted at the transmitter and curtailed in the receiver thereby also reducing noise. The respective operations are termed *pre-emphasis* and *de-emphasis*. In the UK the degree of pre-emphasis is 50 µs, but in America it is 75 µs.

FM demodulation   There are several types of detector in use, the most common being the *ratio detector*. A tertiary winding is connected to a centre tap on the last i.f. transformer secondary, across which the frequency deviations produce an a.f. output. Any amplitude modulations of the carrier are bypassed by a large value capacitor connected across the diodes. The circuit is thus self-limiting and is insensitive to interference pulses that are amplitude modulated.

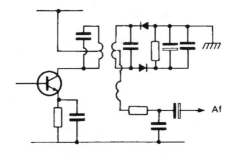

Figure 83. Basic circuit of ratio detector.

Another type is the *Foster-Seeley* discriminator. Unlike the ratio detector, the diodes are connected the same way, and the circuit has a lower distortion. It is not self-limiting, so other means of rejecting amplitude modulation must be employed. One method is to operate the final i.f. stage at saturation level so that it does not respond to amplitude variations: another is to include a separate limiting stage.

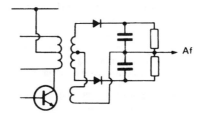

Figure 84. Basic circuit of Foster-Seeley discriminator.

The *phase-locked-loop* is another method of demodulation. A voltage controlled oscillator runs at the centre frequency of the received i.f. signal. A phase comparator compares the oscillator frequency and i.f., and produces a voltage proportional to their difference. Thus when the signal deviates in frequency, a corresponding voltage is produced. The frequency capture range is determined by a low-pass filter connected to the output.

Advantages of the phase-locked-loop are low noise and distortion and rejection of adjacent strong signals The circuit is commonly contained in a single i.c. with a few peripheral components.

After demodulation. the signal is passed through a de-emphasis filter before being fed to the audio amplifier.

Stereo broadcasting

Compatibility A prime requirement of the stereo signal is that it can be received in mono with non-stereo receivers. This precludes the transmission of the L and R channels as two separate signals.

Adding/subtracting The L and R channels are added to form a combine (L + R) signal, and also subtracted to produce (L − R). The (L + R) frequency modulates the carrier at a deviation rate of up to 15 kHz, which is the highest audio frequency transmitted, and is received as a mono signal on mono receivers.

Subcarrier The (L − R) signal is made to amplitude modulate a 38 kHz sub-carrier. Audio frequencies up to 15 kHz thereby produce sum and difference sidebands extending to 38 − 15 = 23 kHz and 38 + 15 = 53 kHz. These are also frequency modulated on the main carrier but as the lowest (L − R) frequency is 23 kHz and the highest (L + R) is 15 kHz, they do not interfere.

Suppressed carrier While sidebands are necessary to convey the (L − R) signal, the subcarrier itself is not essential, but requires substantial transmitter power to radiate. It is therefore notched out of the (L − R) signal along with low frequency sidebands representing the lowest audio frequencies, before it is frequency modulated on the main carrier.

Pilot tone A signal of the same frequency and phase as the suppressed subcarrier is necessary to separate the L and R channels in the receiver. A local oscillator can supply the frequency, but it must be synchronised to the subcarrier to be correctly phased. This is achieved by the transmission of a pilot tone at 19 kHz which is halfway between the highest (L + R) frequency of 15 kHz and the lowest (L − R) sideband of 23 kHz, thus occupying a free part of the spectrum. The 38 kHz subcarrier is obtained at the transmitter by doubling the 19 kHz frequency, so its phase is the same as the subcarrier and it can thus be used to synchronise the local 38 kHz oscillator.

Another function of the pilot tone is to identify a stereo transmission. It is used in the receiver to switch on an indicator called a *stereo beacon*, which remains on as long as the pilot tone is received.

Modulation proportions The f.m. carrier modulation proportions are: (L + R), 45%; (L − R) upper sideband, 22.5%; (L − R) lower sideband, 22.5%; pilot tone, 10%.

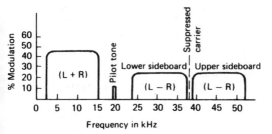

Figure 85. Modulation and frequency spectrum of stereo signal.

Stereo decoding

Most decoders now consist of an i.c. some of which also include the f.m. demodulator. The following descriptions are based on discrete circuits which illustrate the principles although these may be carried out in a different way in the i.c. They may also be encountered in older equipment.

De-emphasis filter   Normally appearing immediately after the demodulator, the filter must come after the decoder, one in each stereo channel, otherwise the high frequencies of the (L – R) sidebands would be lost.

Regenerated subcarrier   The pilot tone is separated from the demodulated signal and used to synchronise a 38 kHz oscillator, or it is doubled in frequency and amplified, thereby dispensing with the oscillator. The result is combined with the (L – R) sidebands so that the original subcarrier is reconstituted. It now takes the form whereby the L channel is modulated on the positive half-waves, and the R on the negative.

Figure 86. Regenerated subcarrier with the L channel modulated on the positive and the R on the negative half cycles.

Synchronous detection   To separate the two channels, a pair of synchronous detectors are required, one switched on by the 38 kHz oscillator to respond to the positive and the other to the negative half cycles.

One method is to connect four diodes in a *bridge circuit* to which the combined (L + R) audio signal and the (L – R) subcarrier signal is applied together with a 38 kHz switching signal and a d.c. potential. The diodes are switched on and off alternately thus sampling the signal at successive half cycles, feeding the output to the two stereo channels. When tuned to a mono transmission, the 38 kHz is absent, so the d.c. voltage switches all the diodes on to conduct the (L + R) audio signal to both channels.

Another method uses a pair of transistors to which the (L – R) signal is applied to their emitters and the 38 kHz switching signal in opposite phase to their bases. Each transistor is switched on and off in turn at opposite half cycles of the switching signal. Both methods have many variants.

Matrixing   If (L + R) and (L – R) are added the result is 2L. If they are subtracted the result is 2R. This can be done by means of a matrix circuit thus resolving the two separate components. However, the (L – R) signal must first be obtained by regenerating the subcarrier and demodulating it, as with the alternative synchronous detection.

Pilot tone filter   A 19 kHz tone is present in the output of the decoder which although not audible could beat with the bias oscillator of a tape recorder and produce an audible whistle on recordings. A low-pass filter must be provided to remove this from both channels.

Signal strength   As the (L –R) signal from which stereo is obtained is only half the modulation level of the (L + R) mono signal (22.5% to 45%), the stereo signal is weaker, thereby producing more noise.

Noise   With a poor r.f. signal, the noise in the stereo mode can be up to 12 dB greater than when operating in mono. Even with a strong signal the noise is some 5 dB higher. Most tuners include an automatic switch that is preset to switch to mono when the signal falls below a certain level. A roof or at least a loft aerial is thus recommended for stable stereo. If reception is weak and noisy, manually switching to mono will improve matters.

BBC stereo test transmissions

These are usually transmitted on Radio 3 a few minutes after
the end of the programme on Mondays and Saturdays, but are
subject to variation or cancellation if circumstances require it.

Stereo test transmissions

Time	LH	RH	Level	Phase	Purpose
2 m	250 Hz	440 Hz	0 dB	—	Identify channels; set reference level
4 m	900 Hz	900 Hz	+7 dB	anti	Set subcarrier phase;* check (L − R) distortion
1 m	900 Hz	900 Hz	+7 dB	in	Check (L + R) distortion
1 m	900 Hz	—	+7 dB	—	Check L to R crosstalk
1 m	—	900 Hz	+7 dB	—	Check R to L crosstalk
1 m	60 Hz	—	−4 dB	—	Check LH response and crosstalk at high and low frequencies
	900 Hz	—	−4 dB	—	
	5 kHz	—	−4 dB	—	(sequence repeated)
	10 kHz	—	−4 dB	—	
1 m	—	60 Hz	−4 dB	—	Check RH response and crosstalk at high and low frequencies
	—	900 Hz	−4 dB	—	
	—	5 kHz	−4 dB	—	
	—	10 kHz	−4 dB	—	(sequence repeated)
2 m	—	—	—	—	Check noise with pilot tone and zero modulation

*Adjust *subcarrier phase* for maximum L or R output, then
crosstalk or *separation* control for minimum crosstalk. If there
is no separation control adjust subcarrier phase for minimum
crosstalk on crosstalk tests.

Notes

1 The 0 dB reference is equivalent to 40% modulation before
pre-emphasis is applied. All test tones are with pre-emphasis.

2 Tests with a duration of longer than a minute have
momentary breaks at the one minute intervals.

3 Balance adjustment should not be made using fixed tones as
there can be sound interference patterns in the listening room
which could give misleading results. Programme announcements
are given from a central position, and it is best to use these to
set the balance.

Interference

Electrical interference   The insensitivity of an f.m. receiver to a.m. eliminates most electrical interference, except where there is a weak signal when it can break through. Car ignition noise contains f.m. components and so is the prime source. New vehicles have adequate suppression, but older ones can cause a lot of interference.

Breakthrough   There are several ways whereby another transmission can interfere. Usually, the interfering signal beats with the local oscillator to produce a 10.7 MHz signal which passes through the i.f. stages.

Second channel interference   Two signals will produce an i.f. output from the local oscillator, one above and the other below it by 10.7 MHz. The receiver is normally tuned to the lower one, but a second channel at twice the i.f. (21.4 MHz) higher, will also produce a 10.7 MHz output.

I.f. harmonic   A signal spaced from the local oscillator by half the i.f. (5.35 MHz), either up or down, generates a 5.35 MHz output. The second harmonic of this is 10.7 MHz which thus produces the interference.

Beat interference   The interfering signal is 10.7 MHz above or below the wanted one. They beat to produce 10.7 MHz which carries the modulations of both. In this case the local oscillator has no part in the process.

Oscillator harmonic   Signals spaced 10.7 MHz above or below *twice* the oscillator frequency can produce a 10.7 MHz output from The oscillator's second harmonic. This is a rare occurrence as the signals are so far outside the r.f. stage tuning range as to be considerably attenuated by it. Furthermore there are at present no authorised communications transmissions in this band.

I.f. interference   Interfering signals are of 10.7 MHz, so the oscillator has no effect. There are no legal transmissions on this frequency so the signal would either be from an illegal short-wave transmitter or from the oscillator of a receiver tuned to the 10 MHz band.

Interference sources   Sources can thus be spaced twice (second channel), $1\frac{1}{2}$ times (i.f. harmonic), once (beat interference), and a half (i.f. harmonic) times 10.7 MHz on the high side, and once (beat frequency), on the low side, from the wanted station. So there are 5 possible interfering frequencies for each broadcast station. Two radio-telephone bands are adjacent to the f.m. band of 88–96 MHz: 70–85 MHz and 103–140 MHz. The most likely source is a mobile or strong base station in the higher band causing second channel interference.

Coax stub   The problem can be tackled once the offending frequency has been identified by fitting a parallel coaxial stub to the aerial lead. Coax possesses inductance and capacitance, so a short length connected in parallel with the aerial plug forms a resonant tuned circuit. With a *velocity factor* of unity, the stub length should be a quarter wavelength of the interfering transmission.

The actual velocity factor varies ranging from 0.67 for the solid dielectric type to 0.85 for the semi-airspaced. The actual stub length should be a quarter wavelength multiplied by the velocity factor. As the v.f. may not be known, the stub should be made full quarter wavelength, and gradually shortened half-an-inch at a time until the interference disappears. The free end should not be shortened but left open.

Alignment

AM alignment   Inject a modulated i.f. signal (usually 470 kHz) into the input of the first i.f. stage, using stray capacitive coupling by clipping the generator lead to an insulated wire. Keeping the generator output low, adjust i.f. coils for maximum output on an output meter connected across the speaker, or stagger tune if the manual so instructs.

Check that the pointer is accurate at both ends of the scale. Loop the generator lead around the ferrite aerial rod, tune to 480 metres medium wave and inject a signal of 625 kHz. Adjust oscillator coil core for maximum output. Re-tune to 200 metres and inject 1500 kHz. Adjust trimmer on oscillator tuning capacitor. Repeat for optimum results. Check calibration at mid-scale: if poor, repeat alignment using 400 metres and 750 kHz instead of 480 metres. Check BBC station on 1500 metres long wave. Correct position with long wave trimmer if fitted.

Melt wax on medium wave ferrite rod coil and slide along for maximum output on a broadcast station at mid-scale. Adjust long wave coil for maximum at 1500 metres.

FM alignment   The best of several methods uses a sweep generator with marker, an f.m. signal generator, an oscilloscope and an output meter.

I.F. alignment Feed the sweep trigger frequency into the X input of the scope: set the sweep output to ±300 kHz either side of 10.7 MHz, and apply to the first i.f. stage of the receiver. Connect the a.f. output to the Y scope input. The discriminator coil must be detuned.

The scope beam is deflected horizontally in step with the sweep generator, while variations in response deflect the beam vertically. This traces a double-humped curve which represents the i.f. amplifier response. The marker produces a blip to identify the frequency of any part of the curve. Adjust the i.f. coil cores in turn to produce a curve symmetrically centred about 10.7 MHz, with a bandwidth of 230 kHz at its top.

Next adjust the discriminator coil to produce a symmetrical S curve with the centre zero point at 10.7 MHz. If there is no adjustable discriminator coil, the required S curve is obtained by adjustment of the i.f. coils.

If a sweep generator is not available, an f.m. signal generator can be utilised with the sawtooth from the scope X timebase used to modulate the generator. A sweep is thus produced in step with the scope X deflection. A separate a.m. generator is needed to provide the marker blip.

RF alignment Tune to 87.1 MHz and feed that frequency to the aerial circuit from the modulated f.m. generator. Adjust the oscillator coil, then re-tune receiver and generator to 108.5 MHz. Adjust the oscillator trimmer. Repeat as required. Using the same frequencies, adjust first the r.f. coil and then the trimmer for maximum output on an output meter.

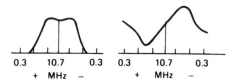

Figure 87. Response curves: (a) I.f.; (b) discriminator S.

F.m. tuner parameters

AM rejection ratio   A 30% tone amplitude modulated r.f.
carrier is injected into the aerial circuit and the resulting output
compared to a 100% f.m. signal. The ratio is a measure of the
ability of the tuner to reject a.m. signals, particularly electrical
interference, although the latter being impulsive may not
always co-relate with the steady tone test. Values of 60 dB are
typical.

Birdie suppression   A transmission on the adjacent channel,
spaced 200 kHz away, can beat with the tuned one in the
receiver circuits to produce a difference signal of 200 kHz. The
reconstituted 38 kHz sub-carrier produces harmonics, of which
the fifth is 190 kHz. These two can beat to generate a 10 kHz
note that warbles with the modulation.

The effect depends largely on the selectivity of the i.f.
stages. Some tuners have a low-pass filter with a turnover
frequency of 53 kHz between the demodulator and stereo
decoder, to reduce it. A suppression figure of 60 dB is average,
though rarely quoted.

Capture effect   If two f.m. transmissions are received on the
same frequency, the stronger is 'captured' and held, while the
weaker disappears. Unlike a.m. it is not heard in the
background. The phase of the two signals invariably differs so
they can be represented by two vectors having different angles.
These are added by joining them and drawing a resultant from
the free ends to form a triangle, the length of the resultant
depicting the amplitude of the combined signal.

The modulations of the interfering signal cause phase
displacements, so the angle of the second vector
correspondingly varies, oscillating around the point of junction
with the first. These oscillations produce mainly *length*
variations of the resultant. Hence the *frequency* modulations of
the weaker signal produce *amplitude* modulations of the
combined signal. As the f.m. demodulator is insensitive to
a.m., the modulations of the weaker signal thus have no effect.

In practice the elimination of the second signal is not total.
It depends on how good the a.m. rejection is. Also i.f. stages
operating at saturation to achieve a.m. limiting can produce
intermodulation effects from the two signals. Furthermore there
is slight phase modulation of the resultant which depends on
the length of the second vector compared to the first, that is the
ratio between the two signals.

The capture effect is therefore specified as the ratio between the two signals which will make the level of the weaker 30 dB lower than that of the stronger. It is usually between 1–3 dB, but down to 0.4 dB has been achieved. Its value is not only in rejecting different transmissions on the same frequency, but multipath reception of the same transmission (ghosting), which produces distortion and stereo impairment.

Frequency response   The transmitted audio range is 30 Hz–15 kHz so the tuner response should be flat over those frequencies. The inclusion of a 19 kHz pilot-tone notch filter may slightly reduce the upper response.

Distortion   Total harmonic distortion depends on modulation level, bandwidth and phase linearity of the i.f. stages, linearity of the detector, the design of the decoder and the inclusion of a 19 kHz notch filter. It is about twice the mono level for stereo. Quoted figures are for 100% modulation, and are generally below 0.2% mono, 0.4% stereo.

IF rejection   This specifies the degree to which a 10.7 MHz signal is rejected. It depends partly on r.f. selectivity, but mostly on the wave-trap, the 10.7 MHz acceptor circuit that shunts the input of the mixer stage. The rejection ratio should be greater than 80 dB.

Limiting   As the aerial signal rises, so does the audio output up to a certain point termed the limiting level. Above that it remains constant and any varying of the signal level has no effect. As the rise to full limiting is gradual, a point 1 dB below it is often quoted. This roughly corresponds to the 30 dB signal/noise ratio level.

Pilot tone rejection   The pilot tone is transmitted at 10% of the maximum modulation level which is 20 dB down. The de-emphasis of 50 μs corresponds to a 15 dB reduction at 19 kHz, so the level of the pilot tone at the output is a total 35 dB below 100% modulation. This is still sufficient to interact with the higher audio frequencies to produce intermodulation distortion, and can beat with the bias oscillator of a tape recorder.

Further attenuation is provided in some tuners by a notch filter tuned to 19 kHz, but it must be sharp enough to have minimal effect on the upper audio frequencies. Rejection should be at least 60 dB. Some circuits use an i.c. which gives total cancellation of the pilot tone.

Second channel rejection   Also called *image rejection*, this depends on the selectivity of the r.f. stage. It should be greater than 50 dB.

Selectivity   The bandwidth of the i.f. stages determines the ability of a receiver to reject a strong adjacent signal when tuned to a weak one. The narrower the bandwidth the greater the selectivity, but this curtails response to large frequency deviations, so creating harmonic distortion on loud signals. Some tuners have switchable selectivity whereby maximum selectivity can be chosen for weak station reception, or maximum fidelity for the strong ones. Alternate channel rejection (spaced at 400 kHz) is typically 55 dB, but ranges from 40 to 100 dB. Adjacent channel rejection (200 kHz) is much less, being from 5 to 12 dB.

The greater the number of variable tuning circuits in the r.f. and mixer stages, the greater the *front-end selectivity*. If this is poor, strong adjacent signals can overload the r.f. or mixer stages producing beat frequencies equal to twice the first frequency minus the second, and twice the second minus the first. This is *third order intermodulation*.

Sensitivity   Noise level diminishes as the aerial signal increases, so the sensitivity rating is that signal which results in a specified signal/noise ratio, usually –50 dB. Sensitivities range from 1–4 μV for mono and about twice that for stereo.

Signal/noise ratio   The measured noise is weighted by passing it through a filter which gives a response close to the subjective effect it has on human hearing (the CCIR/ARM characteristic). The ratio is in comparison with a 100% modulated signal of 1 mV (100 μV). Values of –70 to –85 dB are usual for mono, with about 5 dB less for stereo.

Stereo separation   Up to 25% total harmonic distortion is possible with crosstalk from the opposite channel. 'Spitting' at high frequencies is especially noticeable with poor separation, as well as inferior stereo.

Values of from 40 to 55 dB at 1 kHz are usually specified, but the figures are lower at higher frequencies.

F.m. aerials

Radio wave polarisation   Two separate waves are radiated
from the transmitter aerial, each displaced by a 90° angle from
the other. One is electric and the other magnetic. If the
transmitter aerial is vertical, the plane of the electric wave is
also vertical, while that of the magnetic wave is horizontal. The
magnetic wave has a short range compared to the electric so it
is the electric wave that is used. Receiving aerials must
therefore be in the same plane as that of the transmitter.

Propagation   For long and medium wavelengths propagation
is mainly by a *ground wave* which follows the earth's surface
until it is absorbed. With medium and short waves the principal
radiation is by means of *sky waves* which travel upward to the
ionosphere and are reflected back and forth between it and the
earth's surface. The space between the transmitter and the first
reflection is known as the *skip distance* in which no signals are
receivable. Considerable distances can be covered in this
manner, although the varying nature of the ionosphere makes
reception uncertain and prone to fading.

   At very high frequencies (v.h.f.) a sky wave passes through
the ionosphere and is lost in space. Propagation is therefore by
line-of-sight, although the atmosphere tends to bend the waves
earthward. so the receiving distance is slightly longer than the
visual range. Unusual atmospheric conditions sometimes
increase this distance by greater refraction of the radio waves.

The dipole   A single rod or wire will serve as a receiving
aerial when placed in the field of the transmitting aerial, in the
same plane. The efficiency is high when it is tuned to resonate
at the frequency it is intended to receive by making its length a
quarter of the wavelength of the transmission. If two such
sections are mounted end-to-end with the connections taken
from their junction, it becomes a *half-wavelength dipole* and
the received power is increased. For long wavelengths this may
not be practical and a quarter or eighth wavelengths are more
manageable, though less effective.

Voltage/current distribution   Current induced into the aerial
flows back and forth, and when the length is resonant a
standing wave is set up. The value of the current is at a
minimum at the ends and at a maximum at the centre of the
dipole. The voltage is always out of phase with the current, so
it is at a maximum at the ends and at a minimum at the centre.

Impedance   It follows that the impedance varies over the dipole length, and the central area, having high current and low voltage, is a point of low impedance. With a spacing of 2 cm or so between the inside ends, the impedance is about 75 ohms at this point.

Length   The velocity of the current flowing in the conducting material of the aerial is less than the electromagnetic wave in free space, hence its wavelength is shorter. The true resonant length is therefore less than the calculated value by an amount termed the *velocity factor*.

   This is determined by the ratio of the length to the diameter of the conductor and is normally about 0.95. Increasing the diameter reduces the velocity factor and the length. Aerials for v.h.f. f.m. have to cater for a wide band of 88 to 108 MHz, so a compromise tuning of around 93 MHz is usually chosen giving a length of 1.55 to 1.6 m.

Folded dipole   One way of obtaining a broader bandwidth is to increase the diameter of the aerial and a practical method of doing this without greatly increasing the bulk and weight is the folded dipole. The dipole is formed in a loop having a width of some 3.8 cm. The centre impedance is also increased by up to 4 times which compensates for the decrease which occurs when further elements are added to the aerial.

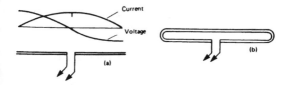

Figure 88. Current and voltage distribution in (a) dipole; (b) folded dipole.

Reflector   When a rod of half a wavelength is mounted behind the dipole, it reflects a signal back to it which reinforces the dipole signal if the spacing between them is a quarter wavelength. The gain over a single dipole is about 3 dB, and the dipole centre impedance is reduced. The aerial also becomes directional as the response to front waves is increased while that to rear waves is decreased.

Director   Another rod, shorter than the dipole and placed
usually less than a quarter wavelength in front, collects signal
and passes it on to the dipole along with its normal pick-up.
This gives a further increase in gain of around 5 dB for the
three elements, greater directivity and lower dipole impedance.
More directors can be added, each slightly shorter than the
previous one, thus increasing the gain up to 10 dB. Although
multi-element arrays are used for u.h.f. television aerials, the
size limits band 2 v.h.f. aerials to three elements, except in
fringe reception areas.

Directivity   The polar response of a vertical dipole is
omnidirectional in the horizontal plane. A horizontal dipole has
a figure-eight response, for which the level for an off axis
signal is cos θ where θ is the angle of incidence. The
response for a multi-element array is a narrow forward lobe.
An interfering transmission is thus rejected if emanating from a
different direction from the wanted one. If it is in the same
general direction but to one side of the wanted transmitter,
rejection can be achieved by aiming the aerial toward the other
side. The wanted signal is reduced slightly being off-centre, but
the interfering transmission is reduced much more as it is
outside the main lobe area.

Cable   coaxial screened cable has a characteristic impedance
given by:

$$Z = \sqrt{\frac{R + 2\pi f L}{G + 2\pi f C}}$$

where R = resistance; G = conductance (leakage) in mhos; L =
inductance; and C = capacitance.

   It is evident that the value of the expression above the line
increases with the length of cable in the same proportion to that
below the line. Hence cable length does not affect the
impedance. For coaxial downloads this is 75 Ω, so there is
maximum transfer of power from aerial to download, and from
download to the 75 Ω receiver input. Some tuners have a
300 Ω input, and there is 300 Ω spaced feeder. All should
match.

Amplifiers

Thermionic valves

Although valves have been long superseded by transistors and ICs for general use, they still have their devotees, and they seem to be growing in popularity for hi-fi amplifiers. Old valve radios are now sought after by collectors and may also appear in the workshop for repair.

Triode The simplest amplifying valve consisting of a cathode heated by a filament, a control grid and an anode. Electron flow from cathode to anode passes through the grid which modulates it according to the voltage applied to it. Although having a lower gain than the tetrode or pentode it has a straighter characteristic and so produces lower third harmonic distortion. It is preferred for hi-fi applications.

Tetrode A finer mesh for the control grid increases its effect but reduces electron flow owing to physical obstruction. To increase the flow a second grid called the *screened grid*, or *beam-forming plates*, is interposed between the control grid and the anode. These plates are connected to a high positive potential and accelerate the electrons towards the anode. A finer mesh can thus be used and higher gain obtained.

Pentode Accelerated electrons reaching the anode bounce off and are attracted back to the positive second grid, so establishing a secondary emission. A third grid which is connected to the cathode, and so is at negative potential, is placed between the second grid and anode. This is called the *suppressor grid* because it repels the displaced electrons back to the anode. As there is a greater choice of high power tetrodes and pentodes than triodes, they are often used as triodes in hi-fi output stages by connecting the second grid to the anode.

Mutual conductance This describes the anode current change for a change of 1 volt on the grid. One unit is the milliampere/volt (mA/V). As current divided by volts is termed conductance for which the unit is the mho, the constant is also specified in millimhos.

It is mutual because it depends on the interaction between grid voltage and anode current. It is also often referred to as the *slope* of a valve.

Anode resistance   A change of anode voltage produces a change of anode current. Dividing the voltage change by the current change gives a resistance value in ohms. For maximum power transfer the anode load should equal the anode resistance, but lower distortion at some sacrifice of power is obtained by a higher load value, often several times higher.

Amplification factor   This expresses the ratio between the anode voltage change to produce a given anode current change and the grid voltage change to produce the same anode current change. As the factor is volts divided by volts there is no unit, but it is usually represented by μ and termed mu. It depends on grid mesh and spacing from the cathode. Some valves intended for use in gain control circuits have a variable-sized mesh over the length of the grid and are termed *variable-mu* valves.

Grid bias   Having curved lower and upper portions to the grid volts/anode current characteristic, the valve must be operated at the mid-point of the straight section. This is done by applying a negative voltage to the grid on which the signal is superimposed.

Grid leak   With a capacitor coupling to the grid, negative charges develop on the grid side which can block the electron flow until they clear. Blocking and clearing is cyclic and is known as *squegging*. It is avoided by returning the grid to an earthy point via a resistor called a *grid leak*. The coupling capacitor and grid leak form a potential divider which reduces the signal applied to the grid. The loss is greater at low frequencies due to the reactance of the capacitor. If the capacitance is increased l.f. loss is reduced, but the grid leak value must then be lowered to avoid squegging. This also lowers the gain of the previous stage because the series CR impedance is effectively shunting its anode load resistor. The product CR in μF and MΩ should not exceed 0.01.

Auto bias   Grid bias is usually obtained by inserting a resistor in the cathode circuit and returning the grid leak to the negative supply. The voltage drop makes the cathode positive with respect to the grid, hence makes the grid negative. This stabilises the operation as any increase in anode current produces an increase in bias thus reducing the current. Signal variations appear at the cathode end which oppose the applied signal between grid and cathode and so reduce gain. A large value capacitor is shunted across the resistor to avoid this, but it is sometimes omitted to produce negative feedback.

Grid current   If the bias is insufficient or the applied signal too great, positive half-cycles drive the grid positive. It then passes current which, because of the impedance in the grid circuit, produces a voltage drop. So positive half-cycles are reduced in amplitude compared to the negative thus generating severe harmonic distortion.

Screen dissipation   Being positive, the screen draws current from the cathode. The current should not exceed the specified value and a series resistor is often included to reduce the voltage. If fitted it should be bypassed with a decoupling capacitor.

Characteristics   Anode current is plotted against anode voltage for a number of curves, each denoting a particular grid voltage. Anode current can also be plotted against grid voltage for a number of curves which give various anode voltages. A *lumped characteristic* is sometimes given which holds good over the working range. With this, anode current is plotted against total voltage and $\mu(V_a + \mu V_g)$. To find the anode current for particular anode and grid voltages, multiply the grid voltage by the mu, subtract from the anode voltage and plot the result against the curve. The lumped characteristic is virtually the same as the anode current/anode voltage characteristic at $V_g = 0$, so that it can be used in the same manner to find currents at other grid voltages.

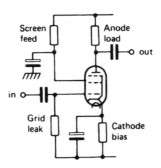

Figure 89. Pentode-amplifying stage illustrating valve principles.

Valve circuits

Stage coupling   Valves require a large anode-to-cathode voltage, this together with an approximately equal voltage drop over the anode load, plus a drop over series decoupling resistor, necessitates a supply voltage in the region of 200V. It is therefore not usually practical to directly couple stages as is done with transistor amplifiers, as this would split the available supply between the stages so coupled.

Capacitive coupling is therefore used from the anode to the grid of the following stage. Often a resistor called a grid-stopper is interposed between the capacitor and grid to prevent parasitic oscillation. Capacitive coupling produces a delay between successive stages, hence a phase shift over the complete amplifier, which varies with frequency. This creates problems when negative feedback is applied, as it can become positive at high frequencies.

Triode stages   Triodes are used in preference to pentodes in hi-fi amplifiers because of their low third-harmonic distortion, but often large output pentodes are used with their screened grids connected to their anodes, which gives a similar effect to a triode, but with higher power. Double triodes, that is valves having two separate triodes of the same type in the same envelope, are often employed. Some that are designed to work at a lower anode voltage can be directly coupled, but this can only be done in earlier stages where the signal level is low, as the low anode voltage reduces the grid voltage signal swing that it can handle.

Output transformer   Being high impedance devices, valves cannot drive a low-impedance loudspeaker directly. Attempts have been made to do so with special output stage configurations or high-impedance speaker coils, but these have met with little success. The standard method is to use a transformer. The primary is connected in series with the output valve anode, while the secondary drives the loudspeaker. The turns ratio depends on the impedance of the loudspeaker which is now virtually standardised at 8 Ω, (Z_2) and the optimum load of the output valve[s] (Z_1). This is somewhat higher than the anode resistance. The ratio is given by:

$$TR = \sqrt{\frac{Z_1}{Z_2}}$$

Distortion is introduced by the output transformer for several reasons. The core can become magnetically saturated at high volume levels, and in the case of single-ended output, also by the d.c. current flowing through the output valve. The inductance does not offer a constant impedance nor winding coupling at all frequencies so there is frequency distortion producing loss of bass. Self-capacitance between turns of the windings is another cause, but this reduces the treble frequencies.

Hi-fi output transformers are designed to minimise these factors but cannot do so totally, hence the efforts to try to dispense with them. However, if this seems a major minus point against valve amplifiers, remember that the speaker coupling capacitor of the transistor variety, is not without its faults either.

Push-pull output stage This consists of two valves in antiphase driving the loudspeaker in concert — one pushes as the other pulls. It has many advantages. Second harmonic distortion generated by a single valve is reduced by some 95%, and third harmonic reduction by around 75%. More power can therefore be obtained with an acceptable distortion level.

Another advantage is that the d.c. current flows from the output transformer primary centre-tap in opposite directions through the winding. The d.c. magnetic field produced by one half of the winding is thus cancelled by that produced by the other, and so there is no core d.c. magnetisation, hence can be no d.c. saturation. Signal current increase in one output valve is matched by a simultaneous decrease in the other, so the supply current remains steady. This applies only to class A stages, but these are always used for valve hi-fi amplifiers.

Ultra-linear output This interesting circuit came into vogue just before the advent of transistors so was never fully developed. It used pentode output valves with the screened-grid connected to a tap on the output transformer primary. Thus negative feedback was introduced which helped reduce the distortions generated by the pentode. It was a sort of halfway stage between a triode and a pentode, giving some of the gain of the latter with some of the distortion reduction of the former.

Driver stage There is no reason why the advantages of the push-pull circuit should not also be applied to the driver stage. Unlike transistor amplifiers, the driver does not have to supply power, only voltage amplification. So with no heat dissipation problems the double-triode can be used here. Capacitor coupling is used to the output stage and from the preceding phase-splitter.

Phase-splitter A transformer with a split secondary winding is one type of phase-splitter, but as transformers introduce distortion, a superior method is to use a triode with equal cathode and anode loads. The signal at the unbypassed cathode of a valve is in opposite phase to that at the anode, and is equal to it providing the loads are equal. Anode and cathode are coupled via capacitors to the grids of the push-pull drivers.

The cathode is thus at about half the supply potential and the grid must be just a few volts below it. This enables a direct coupling to be made to the anode of the preceding stage. Both phase-splitter and preamplifier can also be provided by a double triode.

Servicing valve amplifiers

Many younger engineers will be unfamiliar with valve equipment so here are some basic facts which may help. For servicing, valve amplifiers have advantages and disadvantages over transistors.

Safety The main disadvantage is the high voltage encountered, usually around 200V. When taking test readings with a meter probe, it is recommended that the free hand be put in a pocket. If it is touching the chassis when the other hand accidentally contacts an high tension (h.t.) point, a very unpleasant or even fatal shock could be received.

Especially dangerous is the output from the mains transformer which for a medium-sized amplifier and full-wave rectification, typically has a 350V–0–350V h.t. secondary, thus delivering 700V rms. This is applied to the two anodes of the rectifier valve, so the valve socket bearing their terminals must be treated with respect!

When the amplifier is switched off, there is little danger from the charge in the reservoir and smoothing capacitors as these soon discharge through the output valves which are still hot and conductive. However, test capacitors used to bridge h.t. points for hum or decoupling tests will contain a high voltage charge when removed. These should not be left lying around in this condition, but be discharged after use.

The other safety point to watch out for is that valves get very hot, especially output valves, and they remain so for some while after they are switched off. Paradoxically, it is not the heater that makes a valve hot but heating of the anode by the anode current. This provides a useful test as we shall see. A duster or some other protection should be used to withdraw a valve from its socket to avoid burnt fingers.

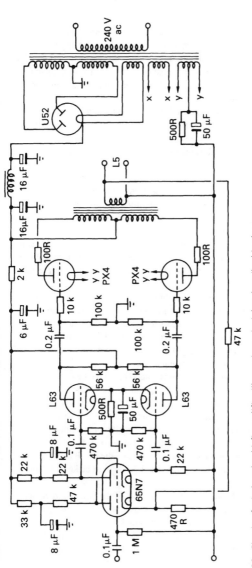

Figure 90. Circuit of triode valve hi-fi amplifier illustrating principles described in the text.

Hum Hum can be a problem with valve equipment. The heaters are supplied with a.c. from a low-voltage transformer winding; the current can be several amps. As electro-magnetic field strength depends on current, and as heater wiring goes to every valveholder, there is the possibility of strong hum fields throughout the amplifier. To reduce this to the lowest possible level, all heater wiring must be tightly twisted right up to the heater pins on each valve holder, thus achieving hum field cancellation.

Hum can be induced from the heater to the cathode inside the valve, as the cathode is at earth potential respecting the signal, whereas the heaters are floating. The effect can be reduced by balancing the heater around earth potential. This is done by either shunting the heater winding with a series pair of low-value resistors having their junction connected to earth, or providing a centre tap on the heater winding which is then connected to earth. This configuration is also used to bias directly-heated output valves in which the heater or filament is also the cathode.

A major source of hum is the mains transformer; the hum fields it produces are much greater than a transistor amplifier transformer. One reason for this is the current flowing in the heater supply winding and the separate rectifier heater winding. Also the h.t. supply windings, although providing lower current than that for a transistor amplifier, have a larger number of turns, and magnetic field strength generated by a transformer is proportional to both current and turns. The transformer must therefore be carefully positioned so that the minimum external field exists around the input circuits.

The problem of hum is compounded by the fact that unlike bi-polar transistors, valve input circuits are of high impedance. In fact, when operating in class A, the grid takes no current at all. High-impedance circuits are more prone to hum pick-up from stray fields than low ones. So all input circuit leads to input sockets, and to volume and tone controls must be screened throughout their length, and the braiding earthed at one end only. Other grid-circuit leads must be kept as short as possible.

Tag boards were the precursors of the printed circuit board and aided assembly by having rows of components neatly laid out in line with their tags wired to the valveholder pins. However, the long grid leads and close proximity parallel-mounted components, were an invitation to hum pick-up and other unwanted couplings. Tag boards should therefore not be used for hi-fi equipment where low hum is a priority.

The high tension supply needs to be as ripple-free as possible. A solid-state voltage regulator such as often used in transistor amplifiers to smooth the supply to early stages, cannot be used with valve amplifiers. Apart from it not being a valve — hence possibly unacceptable to the valve enthusiasts — regulators are designed for much lower voltages than the valve h.t. supply. Capacitor smoothing must be employed in addition to the usual reservoir capacitor. Values are much lower than used with transistor circuits. A resistor can be used for smoothing, but an inductor is much more effective as it offers a high impedance to the a.c. ripple component of the supply, but low resistance to the d.c. current.

Even with all the above precautions and anti-hum measures, it is virtually impossible to totally eliminate hum from a valve amplifier especially when used with speakers having a good bass response. Residual hum will be heard with the volume control full up and no input. However, with good design it can be reduced to almost inaudible levels with normal volume settings and input signals. If it is audible in a hi-fi amplifier under normal operating conditions, there must be a fault.

The most likely is the main smoothing capacitor which can be checked initially by simply bridging another across it. If the h.t. voltage is low, the fault may be in the reservoir capacitor; alternatively one half of the full-wave rectifier valve may be low emission, thereby delivering unequal current on alternate half-cycles. A heater-to-cathode leakage in any valve is a likely cause of hum. Another, rather expensive but less likely possibility, is that half the transformer secondary winding has gone open-circuit so giving half-wave rectification instead of full-wave.

Voltage checks   Fault location in valve equipment is much easier than with transistors. As each stage is capacitor-coupled rather than direct-coupled to the next as with transistor amplifiers, abnormal voltages are confined to that stage and do not produce a chain reaction. Most faults can be rapidly found by means of voltage tests.

Anode voltage should be about half the h.t. supply except on the output stage where it should be higher. No voltage indicates an open-circuit anode resistor or output transformer. A voltage near or equal to the supply shows that no current is being passed; this could be due to the valve, or lack of screen voltage, or an open-circuit cathode resistor.

The latter fault can be detected by a cathode voltage which is near that of the anode. No screen voltage can be due to an open-circuit screen feed resistor or a short-circuit screen decoupling capacitor. A resistance check with the amplifier switched off will reveal which.

Low anode voltage can be the result of the anode resistor going high in value — a not uncommon fault, or excessive current due to positive voltage on the control grid. This in turn is usually caused by a leaky coupling capacitor from the preceding stage, in which case the anode voltage of that stage is likely to be low too.

Another cause is an internal leak between the control and screened grids which effectively puts positive voltage on the control grid. The audible symptom for this is distortion, as the valve will be operating on the high curve of its characteristic.

Visual checks Some quick checks can be made by observing the valves and their operating conditions. Any that does not light up obviously has an open-circuit heater. After a few minutes running, a cautious feel of the glass envelopes will show if any is not passing anode current. A cold or cool valve compared to the others is a possible sign. Remember though that input valves pass less current than those in the output stage and so run cooler. The two (or four) output valves should be equally warm, but do not try this after they have been running for more than a few minutes as they will be too hot to touch!

A valve may feel excessively warm or be seen to be overheating by grids glowing, or the anode glowing a dull cherry red. An internal leak between grids or leaky coupling capacitor is the most likely cause. A short-circuit cathode bypass capacitor is also a possibility as this removes the bias. (Note that with valves, bias inhibits the current, not enables it, as with transistors.) Unlike bipolar transistors that quickly self-destruct by thermal runaway, valves will often be found to be little the worse for such an experience.

Disturbance testing This is a very useful method of fault location. With transistor amplifiers a fault in one stage can upset every other because the direct coupling can affect the d.c. operating conditions in a chain reaction. With valve amplifiers each stage is more or less isolated as regards d.c. conditions, so all stages are likely to be fully operational except the one having the fault. A logical stage-by-stage diagnosis can thus be employed, starting at the output stage. First, voltage checks are made in each stage, then the anode and grid pins are scratched with the meter probe. This generates a wide band of random frequencies which should produce noise from the loudspeakers. It should get louder, progressing from output to input stages. A more elegant but not necessarily more effective way is to use an audio signal generator. These instruments come into their own though for more precise circuit testing.

Instability   The commonest form with valve amplifiers is *motor-boating*. The term is a good description of the symptom. The fault is really low-frequency instability and the cause is likely to be the main smoothing capacitor or one of the supply decoupling capacitors. Whistles or hoots are more likely to be due to an open-circuit screen decoupling capacitor or anode decoupling. The latter when provided, is connected to the junction of the top end of the anode load resistor to another resistor which goes to the supply.

Squegging   This effect is not unlike motor-boating but the cause is quite different. The signal is alternately cut off then restored at regular intervals, usually two or three times a second. It is due to electrons accumulating on the control grid of a valve until a negative charge is built up which blocks the anode current. As no more electrons arrive, and those on the grid leak away, the charge soon disappears and anode current is permitted to flow again. Another charge builds up and the action is repeated. The cause is the lack of a d.c. path from the grid to earth that will allow electrons to leak away before a charge can built up. The path is normally provided by a high-value grid-leak resistor connected from grid to earth, and this is likely to be open-circuit.

Loss of gain   Low supply voltage is a possible cause of loss of gain which in turn can be due to a low-emission rectifier or defective reservoir capacitor. Any valve could cause it if it has low emission or low conductance, including the output pair, but if only one of the pair is low or even completely non-conducting, it often has little effect on the audible output level. It seems that the resulting higher supply voltage and other factors make the remaining valve work harder.

Another common loss of gain is an open-circuit cathode-bypass capacitor. Without it, a signal is developed over the cathode bias resistor which is effectively in series with the grid-to-cathode input signal, but is in opposite phase. It is thus a form of negative feedback, which reduces gain. The capacitor bypasses this antiphase signal and so allows the full input signal to be applied. Capacitor failure thus causes loss of gain.

Valve testing   Unlike transistors, valves deteriorate with use which means that the performance of an amplifier gradually falls off. Often there is no audible effect but for the enthusiast this is not good enough, everything must be perfect; so valves should be tested when the equipment is serviced, and any that are below par, replaced.

The minimum test applied by a tester is for emission and inter-electrode short-circuits. These were the tests applied by the old Mullard card tester which was intended for operation by semi-technical staff. However a more complete test as applied by the Avo tester, Taylor, and others, measured in addition the conductance in milliamps-per-volt. It worked by measuring the anode current which thereby checked the emission, backing it off to zero, then applying 1 V to the grid and noting the difference in anode current.

As valves were made with many different bases and pin configurations, a panel of different valve holders is an essential part of any valve tester. And as connections to even the same type of holder varies from one type of valve to another, a multi-way switch is needed to enable the correct connections to be set up. Alternatively, a larger panel of holders with several of the same type but having different connections for the various valve types could be used. The correct electrode voltages for each type also need to be set, which can be done by switched controls. All these appeared on the old testers and a book was supplied with each instrument giving the pin connections and voltage settings for most known valves. With the Mullard tester all this was set up by inserting a punched card for each valve type.

Output valves should be matched to within close limits, for both emission and conductance, to get a balanced output signal.

Vintage radios

There is a growing interest in vintage radios. Apart from the reproduction cabinets with modern electronics that are widely advertised, genuine models which change hands at quite high prices are collected and prized. They may therefore appear in the audio engineer's workshop, so the following information could be useful. As these use valves, they are included in this section rather than in the preceding radio chapter.

The most sought after are the pre-war models up to 1939. These were quite different in many respects from those that followed the war from 1945 onward. During the war the 'civilian receiver' was produced which was very basic having a four-valve circuit and medium-wave reception only. The cabinet was of unveneered wood, and there was no manufacturer's identification, they all made the same model; but it was sometimes possible to identify the maker by the components. After the war long-wave conversion units were sold to enable the *Light programme* to be received on the long wave band at 1500 m.

There were three formats: the table radio; the console radio in which the same radio chassis was used in a floor-standing cabinet, and the radiogram in which a large wide, floorstanding cabinet with a lid housed a similar chassis but with an autochanger record playing unit. Better models had a larger loudspeaker and output stage.

Pre-war valves Those commonly used were known as *English* valves which had either a 5-pin or 7-pin base. Connections to the 5-pin base were standard, but there were variations with those of the 7-pin variety. They all had the same heater connections, and the heater voltage was a standard 4.0 V except those designed for a.c./d.c. operation, or for battery use in which case they were 2.0 V.

Most valves except output valves, had a top cap connection, and this was often an unpleasant trap for the unwary. The common connection to the top cap was the control grid. And a quick check could be made by putting a finger on the top cap to produce a healthy hum from the audio stages or noise from the radio ones. However, with some radio valves the top cap was connected to the anode and had up to 200 V on it. Unwary fingers often suffered! With Mullard valves, the top-cap grid valves had the suffix B, while the top-cap anode valves had the suffix A, but with a valve surrounded by coil cans it was not always easy to see the type number.

With 5-pin output pentodes, there was usually a side screw connection on the valve base, and this was connected to the screened grid.

Most r.f. valves had a metallised coating which served as screening.

This was connected via a turn of copper wire at the top edge of the base, down through the inside of the base to the cathode pin. It often happened that this wire ceased to make contact with the metallising which thereby was not earthed and so resulted in instability. The valve could frequently be saved

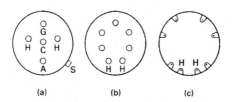

Figure 91 (a) English 5-pin base with standard connections, used for triodes, or pentodes with a side-connection for the screened grid (b) English 7-pin base for pentodes, heptodes and so on. No standard connection other than the heater as shown (c) side-contact base with heater connection shown

by wrapping a few turns of copper wire around the bottom of the metallising and soldering the turns together, then soldering it to the existing wire or, if this had broken off, externally to the cathode pin.

A less common type of valve base was known as the P-base, believed to be of continental origin; it had side contacts arrayed around the base. The holder was a huge well-like socket into which the whole valve base was inserted, and had spring contacts around its internal circumference. They took up a lot of room and were very troublesome.

A.c./d.c. circuits   D.c. mains was common in many parts of the country so some models were designed to operate either on d.c. or a.c. supplies. No mains transformer was possible, but the h.t. supply was taken directly from the mains via a half-wave rectifier. The heater supply was the problem. Special valves having high voltage heaters from 13–55 V were used and all connected in series. When added, the combined heater voltage was still lower than the mains voltage, of which there were three standards, 200 V, 210 V and 230 V. The difference was made up by a mains dropper, a large wire-wound resistor that had tappings for the three supply voltages, and which got very hot.

An alternative was the line-cord dropper. This consisted of a mains lead containing resistance wire. This too often had tappings resulting in a plethora of wires emanating at the receiver end. An open-circuit was a nightmare to trace and repair. Fortunately these were less common. One characteristic that all a.c./d.c. sets manifested when used on a.c., was hum.

R.f. circuits   Pre-war r.f. circuits were quite complicated and no two were alike, differing widely between models of the same make. There were a few t.r.f.s, but mostly they were superhets. Stagger-tuning was often used to achieve bandwidth in the i.f.s; the i.f. frequency differed between models but was normally around 110 kHz. R.f tuning also often had stagger-tuning.

The frequency changer was usually a heptode having two more grids than the pentode. These were connected to function as the local oscillator. The r.f. input from the aerial circuits was injected at the control grid, so the whole valve served as r.f. amplifier, oscillator and mixer. A signal at i.f. frequency appeared at the anode, the anode load being the primary of the first i.f. transformer.

Some frequency changers were of the triode-hexode type although they became more common in post-war models. With these, the triode operated as a separate oscillator; while the hexode had one extra grid (compared to a pentode) internally connected to the grid of the triode. The oscillator output was thus injected into the hexode mixer.

The load of the i.f. valve was the primary of the second i.f. transformer; the four windings of these transformers providing the required selectivity. Demodulation was with a diode which usually was part of a triple-function valve, a double-diode-triode. The second diode rectified the carrier at the i.f. valve anode to provide what was called AVC (automatic volume control) although this was really automatic gain control.

A.f. circuits Output from the signal diode was applied to the volume control through a simple RC filter to remove r.f. and from there to the grid of the triode which served as the first a.f. stage. The triode anode was capacitively coupled to the grid of the output valve. This was matched by a transformer to a 3 Ω loudspeaker.

Energized loudspeakers These were commonly used at the time. Instead of a permanent magnet, the loudspeaker had a large coil which was connected in series with the h.t. supply between the reservoir and smoothing capacitors.

It thus served a dual role, to energize the loudspeaker and to provide inductive h.t. smoothing. Another cause for caution with 200 V at the loudspeaker!

As the field was produced by unsmoothed h.t., this would be expected to produce hum. However, a small 'hum-bucking coil' was connected in series with the speech coil; this picked up hum from the field coil, but being connected in antiphase, it cancelled the effect of field coil hum.

When connected the wrong way round as they sometimes were, this made the hum worse.

Economy and luxury The above description was for the normal 5-valve radio, the rectifier being counted as a valve. Economy 4-valve models used a double-diode-pentode instead of a double-diode-triode plus output pentode. So the triode a.f. amplifier was sacrificed. Luxury 6-valve models had an r.f. amplifier before the frequency changer and usually had one or more short-wave bands. Others, especially radiograms had the normal radio valve complement, but had a push-pull output stage, but these were less common.

Magic-eye tuning An extra often seen on luxury models was the magic eye. This was really a small cathode-ray tube with built-in triode amplifier. The green screen was a circular disc about $^3/_4$ inch in diameter, that was viewed from the end of the valve. It was fed from the AVC line and produced a shadow which closed when a station was fully on tune.

Construction Conventional wiring was mostly used but
there were some notable departures. Early Philips models were
wired throughout with single bare uninsulated wire that was
bent at angles to avoid other wires. Small solder-filled wire
spirals were used as connectors. Another horror from the
Philips stable was the wet electrolytic. It worked well enough
until the chassis was upended for servicing — then disaster.
Yet another from the same source was wiring run everywhere
in harnesses. Some was half coloured so that when it was
twisted it came out of the twist a different colour than it went
in. Tracing leads was a nightmare. This same wiring was
covered with a rubber insulation that perished in a few years.
So the harness became full of bare wires all short-circuiting.

Another brilliant innovation could be found in the HMV/
Marconi brands. All small components were contained in a
metal case that was sealed with pitch. Rows of terminals
appeared on the outside to which connection were made.
Servicing was virtually impossible without a service manual.
Even then as many components were internally connected, it
was difficult to isolate faulty ones.

Large tag boards with components underneath and
unreachable without disconnecting and removing the whole
board, as well as on top, also made life difficult with some
models.

Post-war models These were generally simpler and much
easier to service.

Layout was better with few of the quirks found in pre-war
models. Valves were mostly standardised with the international
octal base which had eight equally-spaced pins around a central
locating spigot. Connections were also standard except for a
few special types.

Heater voltage was 6.3 V, and while some r.f. types were
metallised, the generally smaller size made metal screening
cans the preferred screening option. Anode top-caps made a
welcome disappearance! To start with, a single range was used
in almost every receiver; these were, with their equivalents:

Frequency changer	l.f.	Double-diode-triode	Output
ECH 35; OM 10;	EF 39; OM 5;	EBC 33; OM 4;	EL 33; 6AG6
X61M; X 147	W 147	DH 63; DH 147	

Rectifiers also used the octal base, but their heaters were a
standard 5 V, with one or two exceptions.

Later the all-glass type of valve appeared as well as the
Mazda octal base, and bases and connections diversified again.

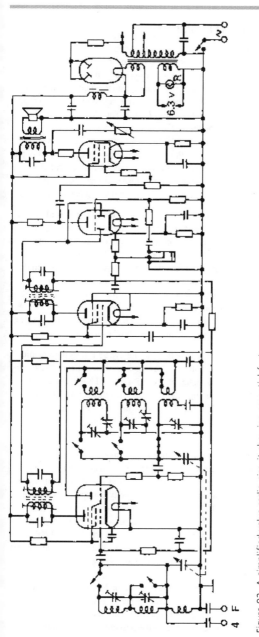

Figure 92. A simplified valve radio circuit showing essential features.

Audio transistors

Point contact   The earliest device known as a *transfer resistor* (transistor) consisted of two fine metal points in close proximity on a semiconductor material having a deficiency of free electrons (P material). Current injected into the material via the one point, termed the *emitter*, released a larger current gathered by the other, the *collector*. The closer together the points were, the larger the current gain.

A later version, easier to manufacture, had cavities etched on opposite sides of the germanium base material with tiny jets of hydrofluoric acid into which the metal contacts fitted. Spacing was thus the thickness of the material between the cavities. Disadvantage was vulnerability to mechanical shock.

Junction transistor   A further development diffused indium into opposite side of an N-type doped germanium base by heat. These diffusions developed P-type impurities at their interface with the germanium, thus producing a PNP sandwich. The deeper the diffusions the thinner the central N portion and the higher the gain. But if taken too far they touch and short-circuit. Excessive heat generated during operation causes the diffusion to continue and results in a short-circuited device. The diffusion is not precise enough to obtain a narrow spacing of close tolerance, so wide spreads of gain result.

Alloy diffused transistor   In this type a P-doped germanium slice forms the collector, having an emitter pellet diffused into its surface. The pellet has both P-type and N-type additives, but the N-type penetrate more deeply than the P-type. They are thus sandwiched between P-type materials, and they merge with N-type diffusions from an adjacent N-type pellet which serves as the base connection.

Planar transistor   A slice of pure monocrystalline silicon is doped with either a P-type impurity such as boron, or N-type such as arsenic. It is oxidised on one side and a hole etched in the oxide. Through this the opposite type impurity is diffused to form a well. The wafer is re-oxidised and a smaller hole etched. Impurity of the original type is diffused through this to form a smaller well within the previous one. Thus a PNP or NPN sandwich is formed. The centre 'filling' can be made very thin by careful control of diffusion, so large gains are obtainable.

A large number of units can be made simultaneously on the same slice so reducing manufacturing costs considerably. Furthermore, generated heat causes both wells to diffuse in the same direction, so reducing the possibility of short-circuiting.

Epitaxial planar transistor   This is similar to the planar type but a P-type layer is grown on to the surface of the N-type wafer before the oxide layer is deposited. This serves as the base, so only a single diffusing process is required to form the emitter. Most small transistors are of this type because they are easier to make, have higher breakdown voltage rating, higher gain, and slightly lower noise. They are made in both PNP and NPN versions.

Polarity   Theoretically, the transistor sandwich could be reversed and the emitter and collector connections interchanged. However, the collector is made much larger than the emitter to readily attract current carriers from it and so prevent their excessive combination with those of opposite polarity in the base.

FET   Various constructions are used for field effect transistors; with one, P-type beads are either alloyed or diffused on either side of a slice of N-type material similar to a bi-polar transistor. But instead of current passing from one to the other across the slice, it passes down the slice between the beads which control it by an electrostatic field. The material is termed the *channel*, and can be either N- or P-type. The emitting electrode is the *source:* the control, the *gate:* and the final electrode, the *drain*. The device is thus voltage-controlled like a valve and has a high input impedance.

MOS FET   As there is no need for direct contact between the gate and the channel they are insulated in the MOS FET by a layer of silicon oxide. It thus has an even higher input impedance than the ordinary FET.

VMOS FET   Power is limited when current passes along a narrow channel, so here it is made to pass through the slice like a bipolar transistor, thereby achieving a greater conduction area. The gate is formed into a V cut into one surface which produces a surrounding field in the material.

   Output stages are increasingly using MOS FETs. Their negative temperature coefficient virtually eliminates the risk of thermal runaway. The absence of minority carriers in the control region gives a fast response, hence a good slew rate and the high input impedance simplifies the driver stage.

   A static voltage applied to the gate can puncture the oxide layer and so destroy a MOS FET device Conductive packing materials are used to avoid this. The hand should be discharged to earth before handling and soldering irons earthed when installing.

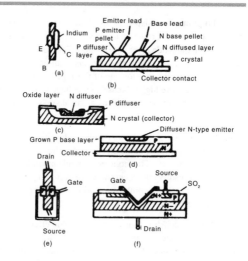

Figure 93. Transistor constructions: (a) junction transistor; (b) alloy diffused transistor; (c) planar transistor; (d) epitaxial planar transistor; (e) FET, (f) VMOS FET.

Figure 94. Transistor symbols: (a) bipolar NPN; (b) bipolar PNP; (c) P-channel FET; (d) N-channel FET; (e) MOS FET.

Transistor characteristics

Admittance (Y) The small signal common-source transfer admittance for an FET, symbol Y_{fs}, denotes the drain current/gate volts. The unit is the Siemens which is equivalent to the mho. It is usually specified in μS or mS and is parallel to the mutual conductance of the valve.

Alpha (α) The current gain of a transistor in the common base mode, now rarely used.

Beta (β) The current gain of a transistor in the common emitter mode, now superseded by h_{fe}, for small signal, and h_{FE} for d.c. gain. The h_{FE} is obtained by dividing the collector current by the base current.

Bias A bipolar transistor must be forward biased in order to conduct. That is, a positive current must be injected into the base for an NPN, or a negative current for a PNP device. With FETs of the enhancement type, forward gate voltage bias is required (positive with N-type and negative with P-type channels). With depletion type, reverse-gate voltage bias is necessary. Most output devices are of the enhancement type and so require bias of the same polarity as the supply, thus simplifying the biasing.

Heat dissipation A major factor especially with bipolar transistors. A temperature rise produced by the power expended in the device reduces the resistance so increasing the current and power, thereby raising the temperature further. Current and temperature continues to rise in what is called *thermal runaway*, until the device short-circuits and is destroyed.

Temperature must therefore be controlled by adequate heat removal. There are three barriers it must cross: first, the thermal resistance between the collector junction and the casing of the device; second, the thermal resistance between the device and its mounting; and third, the resistance between the mounting, which is usually a heat sink, and free air. If the device is free standing, the second and third barriers are combined. Thermal resistance is defined as the temperature difference that exists across the barrier for each watt of power produced and is specified in degrees C/watt. Maximum ambient temperature, heatsink resistance or power can be determined given the others, and the device junction maximum temperature.

The relationship is expressed as:

$$T_a \leq T_j - P\left(\theta_j + \theta_c + \theta_H\right)$$

$$\text{or } \theta_H \leq \frac{T_j - T_a}{P} - \left(\theta_j + \theta_c\right)$$

$$\text{or } P \leq \frac{T_j - T_a}{\theta_j + \theta_c + \theta_H}$$

where T_a is the ambient temperature; T_j is the junction temperature; P is the power in watts, and θ_j, θ_c, θ_H, are the thermal resistances of the junction-to-casing, casing-to-mounting, and heatsink respectively.

Maximum junction temperatures are between 150° and 200° for silicon and 100° for germanium transistors. The following are approximate values of θ_j for various encapsulations, but exact values depend on type:

TO3	0.5–2.5°C/W	TO39	15–36°C/W
TOP3	1.5°C/W	TO66	4.5°C/W
TO5	15–35°C/W	TO220	3–4°C/W

Mica washer resistance between TO3 devices and heatsink is 0.5°C/W, while direct contact between device and mounting is about 0.2°C/W.

H_{FE}/H_{fe} The d.c. current gain and small signal current gain formerly called beta. Large spreads are common up to 700%, so some types are divided into three groups having a suffix A, B, or C, with C having the highest gain. Sometimes these are colour coded with a red (A), green (B) or blue (C) spot. The H_{fe} is usually specified at a single operating frequency which for audio is 1 kHz; it is much less at frequencies near its upper limit (100 kHz for audio). Gain also varies with collector current, being lower at higher currents and often also at very low currents.

Limiting parameters   Data sheets specify maximum voltages and currents for the various electrodes. These are usually for 25°C (77°F), and must be derated for higher ambient temperatures or smaller heatsinks. Uprating is also possible if larger heatsinks are used. For long life and reliability a device should be operated well within its maximum ratings

Basic transistor circuit features

Common base The signal is applied across the emitter base junction, the base being grounded and common to input and output circuits. Output is taken from the collector side of the load. Has voltage gain but current gain is less than unity. Lowest input impedance of about 50 Ω, highest output impedance of around 1 MΩ, 0° phase shift.

Common emitter The signal is applied across the base/ emitter junction; the emitter being grounded and common to input and output circuits. Output is from the collector side of the load. Has voltage and current gain. Input impedance 1–2 kΩ, output impedance about 30 kΩ, 180° phase shift.

Common collector Also called *emitter follower*. Signal is applied across base to ground. Output taken from emitter side of emitter load. Collector is grounded as respects the signal (but not necessarily the supply), and so is common to input and output. Has current gain but voltage gain is less than unity, being:

$$V = \frac{Z_e}{Z_e + \left(\dfrac{1}{g_m} + \dfrac{Z_s}{H_{fe}} \right)}$$

in which Z_e is emitter impedance; Z_s is source impedance; $1/g_m$ is the reciprocal of mutual conductance which depends on internal resistances and emitter current, typically 50 at 1 mA for small transistors, very low for output transistors.

High input impedance is a feature, being approximately equal to the H_{FE} multiplied by the emitter load, shunted by the base bias resistor if used. Output impedance can be very low. There is 0° phase shift.

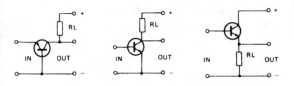

Figure 95. (a) common base; (b) common emitter; (c) common collector.

Class A In order that both half-cycles of the applied signal can be amplified equally and without distortion, the transistor is biased to operate at the mid-point of the straight portion of the transfer characteristic. Excursions of the collector current either side of this point thereby remain in the linear portion. If the bias point is not at the centre, one half-cycle encounters a curve earlier than the other, so the signal-handling capacity is reduced to the shorter section.

For single, common emitter transistors, the device is mid-point biased when the collector voltage is roughly half that of the supply, or the emitter voltage for the common collector mode. This is achieved for the common emitter stage when the value of a base bias resistor, which is returned to the collector, is equal to the collector load resistor multiplied by the H_{FE}.

Base bias can be derived from a tap on a potential divider across the supply if thermal stability is important as with germanium devices, or from a single resistor to the supply rail or collector. The latter gives negative feedback over the stage which reduces gain, but also noise and distortion. With directly coupled circuits, the bias is obtained via another transistor.

A resistor is usually included in the emitter circuit to provide thermal stability and also to even out the effects of H_{FE} spreads. A device with a higher H_{FE} than designed for has a higher collector current for the given bias. The result is a larger voltage drop over the emitter resistor which reduces the base/emitter voltage and with it the collector current. The resistor is bypassed with a high-value capacitor to avoid negative feedback of the signal.

For *output circuits*, class A is only about 20% efficient which means that for large output powers, considerable heat must be dissipated. The *push-pull circuit* is universal for all but very small output stages. In this, the current increases through one transistor as it decreases in the other, both being biased to their mid-points. Cancellation of second harmonic distortion is thereby achieved and about two and a half times the power of a single transistor is obtained.

If the same type of transistor is used for both units, the phase must be split so that one base is driven positive while the other is driven negative. It is more usual for opposite polarity transistors to be used, an NPN and a PNP, or a P channel and N channel if they are FETs. Then a signal of the same phase can be applied to both. This is termed a *complementary pair*.

Class B To improve efficiency and reduce heat generation, the transistors are biased to the bottom of their characteristic which in true class B is zero bias. Each unit then handles only one half-cycle. In practice there is a displacement where the two halves join, and this generates *cross-over distortion* which is mostly third harmonic. As the discontinuity is proportionally larger for small signal amplitudes than large ones, the audible effect is worse at low volumes.

A related effect is *switching distortion*. As each transistor switches off it stores a charge and the corners of the wave are rounded off at the switching point, so that the two halves do not connect.

Class B variants Various techniques have been devised to minimise cross-over and switching distortion, one being to allow a small quiescent current to flow through both transistors. Unless a large quiescent current is passed which is almost a reversion to class A, it does not greatly reduce the distortion.

The output current is compared with a constant current source and a variable bias is produced which controls the transistors alternately during the half-cycle they are normally switched off. During large excursions they are biased off as with conventional class B, but for low amplitudes they are fully on and so operate in class A. For intermediate levels they are progressively biased between the two extremes. Different makers arrange the biasing circuit variously and give their own names to the resulting circuit, but the basic principle is the same.

Other forms of distortion in the output stage are: *transfer characteristic non-linearity, mismatch in H_{fe}, mismatch in cut-off frequencies* and *mismatch in input impedances*.

Transfer characteristic non-linearity The 'straight' part of the characteristic curve is not perfectly linear, so output does not correspond exactly with input and distortion is generated. This is mostly third harmonic and can be calculated from its curve as follows.

Determine from the curve the base current I_b producing the peak collector current I_{c1}. Then find the collector current I_{c2} which corresponds to $0.5I_b$. Calculate:

$$D = \frac{i_{c1} \div I_{c2} - 2}{2(I_{c1} \div I_{c2} + 1)} \times 100\%$$

Mismatch in H_{fe} Mismatched output transistors produce unequal half-cycles and thereby second harmonic distortion. The amount can be determined from the ratio of the two H_{fe}s:

Ratio H_{fe}/H_{fe2}	1.1	1.2	1.3	1.4	1.5	1.6	1.7	1.8	1.9	2.0
Distortion %	2.4	4.6	6.5	8.3	9.8	11.5	13.0	14.5	15.5	16.5

Mismatch in cut-off frequencies As signal frequencies increase, phase shift is introduced by the transistor, which gets more severe as it approaches cut-off point. An output pair having different cut-off points, even if above the upper audio frequencies, will manifest different phase shifts in the high frequencies and produce intermodulation distortion. The effect is controlled by using transistors having a high cut-off point and limiting the h.f. response of the amplifier by an input filter.

Mismatch in input impedances   Second harmonic distortion is produced by differences in the source impedance which can arise with push-pull drivers having unequal H_{fe}s. Also, as the input resistance varies with input current it must be swamped by a high source resistance to avoid non-linear distortion.

Total distortion   If distortion levels of individual harmonics are known the total is given by:

$$THD = \sqrt{D_2{}^2 + D_3{}^2 + \dots}$$

Class C   Not used for audio. The transistors have no forward bias so conduct for only the upper part of a half-cycle. A tuned resonant load 'rings' and so fills in the missing part of the waveform. It is an effective amplifier of radio-frequency carriers having an efficiency higher than 90%.

Class D/pulse width modulation   Most of the distortion generated in an amplifier arises in the output and driver stages. This is here avoided by converting the analogue audio signal into a series of pulses at a frequency of 100–500 kHz. Their amplitude and frequency are constant, but their width varies according to the audio. Once converted they are like a digital signal, immune from noise or distortion. Modulation of the pulse width is done after an initial stage of analogue amplification. One method uses a comparator and a clock. The audio is fed to one comparator input and the clock pulse to the other; the duration of the output pulse then depends on the amplitude of the audio signal compared to the clock pulse.

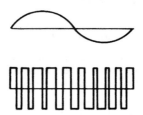

Figure 96. Analogue and corresponding pulse width modulation.

The output transistors are either switched on or off so they dissipate negligible power and can be zero biased. Small transistors can thus produce high power with little heat. Pulses are converted back to analogue by the loudspeaker itself which cannot respond to the high pulse frequency but only to their average value. This average is proportional to the pulse width, so it reproduces the original audio. Although the pulses are too rapid for the cone to follow, they can drive high currents through the coil even in the absence of audio modulation, as there is no back e.m.f. to counter them. A low-pass filter is therefore essential to remove individual pulses from the output.

The problems with class D have been the high-power fast switching speeds required for the output transistors, although V-FETs are capable of them; radio interference generated by the power switching; and frequency, harmonic, and intermodulation distortions produced by the modulator.

Darlington pair/super-alpha pair   Two transistors have their collectors connected and the emitter of the first feeds into the base of the second. The input is taken to the base of the first, and the output from the collector or emitter of the second. The first unit is an emitter follower and a very high input impedance in excess of 1 MΩ is achieved. The gain is equivalent to the individual H_{fe}s multiplied. As power output transistors have low gain, a Darlington output pair in a single encapsulation is often used to give gain with power.

Differential pair/long-tailed pair   Two transistors are connected with a common emitter resistor which stabilises their d.c. operating conditions. Input is connected across the two bases and the output across the two collectors. If one side of the input circuit is earthed, one of the bases can be a.c. coupled to earth. Signals of opposite phase are produced at the collectors, so when these are combined harmonic distortion generated by transfer curvature is cancelled. Supply-ripple, or noise, affect both transistors and so is cancelled. This is termed *common-mode rejection*.

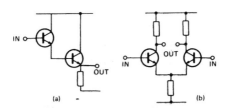

Figure 97. (a) Darlington pair; (b) differential pair.

Driver stage   The drive current required by an output stage is the output current divided by the H_{FE} of the output transistors. As this is usually well below 100, the drive current is quite large and the drive transistors must be power types with heat-sink cooling. The output impedance should be high to obtain a high ratio to the input resistance of the following stage. Then variations of its input resistance will be swamped and have little effect on the applied signal.

The stage may consist of a single transistor, a Darlington coupling with each output transistor, a complementary pair or a differential pair. The driver output is directly coupled to the bases of the output transistors, which have a variable resistor between them to adjust their voltage difference hence the bias. Diodes are also included for temperature compensation.

Feedforward   A fraction of the output of a power amplifier is compared with its input, The difference is the generated noise and distortion, which is amplified by a separate high quality class A amplifier. This is then added in antiphase to the output of the main amplifier, so cancelling the distortion. There are various versions of this principle. One is the *current-dumping amplifier* that uses an auxiliary differential amplifier to drive a high power class B output stage. Part of the power output with its distortion is taken back to the inverting input of the differential amplifier which thus produces a signal with negative distortion. Outputs from both are fed to the load via impedances which, with the amplifiers, form a bridge configuration and the positive and negative distortion values cancel.

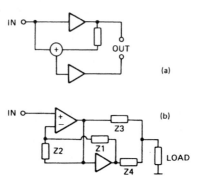

Figure 98. (a) feedforward system; (b) basic current dumping circuit.

Input stages   As these handle small signals, even moderate amounts of noise could be large in proportion. Most of the amplifier's gain follows, so noise here has a far greater effect than at any other stage. Thus low noise is a major factor. It is achieved with low-noise transistors and resistors, negative feedback and operating at low emitter currents.

Pickup cartridges generate thermal noise which is a function of the resistance of their windings. When related to an average signal level a typical signal/noise ratio for a moving magnet device is 80 dB. For a moving coil, although the resistance is lower, so also is the output and a typical signal/ noise ratio is about 60 dB. To avoid degrading the performance. the signal/noise ratio of the gram input should ideally approach these figures.

Transistor noise arises from two main causes. The *base/ emitter internal resistance*, which generates *thermal noise*, and *emitter current*. As the base/emitter resistance of a power transistor is low, it can be used at low current to achieve a low-noise input stage especially with a low source impedance as with a moving-coil pickup. Base/emitter resistance decreases with increasing emitter current, but noise due to emitter current increases. There is thus an optimum emitter current for any transistor at which the total noise is at a minimum.

To preserve optimum signal/noise ratio, volume controls are always put after the input stages. Noise as well as the signal is thereby reduced as when operating at low volume. But this makes the stage vulnerable to overload owing to a high input signal, especially from the gram pickup. So it must be designed to handle signals considerably higher than normal without excessive distortion. Up to 32 dB overload is a reasonable rating. This can be achieved by increasing the emitter current, which also increases noise. Thus the design is a compromise.

Thermal stability in the input stage is important because with most designs there is direct coupling through to the output stage and a small d.c. drift can cause a large deviation at the output.

Matching   The output impedance of a signal source is effectively in series with the signal. It thus forms a potential divider with the input impedance of the following amplifier. If the two are equal, the signal appearing at the input is half of that generated. The signal/noise ratio is thereby degraded by 6 dB. If the input impedance is ten times that of the source, the available signal is 0.909 of that generated and the signal/noise ratio worsened by only 0.9 dB. So for voltage transfer, the following impedance is usually made ten times that of the source.

With inductive sources, load differences can vary the frequency response. Gram input impedances should match those specified by the cartridge maker. The standard is 47 kΩ, but other values are also used. Negative feedback is customarily applied to the emitter of the gram input stage, but as the emitter circuit is also part of the pickup circuit, its reactance and that of the lead capacitance can modify the negative feedback, so producing frequency response anomalies. This is especially so if the feedback is part of the RIAA equalising circuit. To avoid this, the first stage should be a buffer between input and equalising.

For optimum power transfer the impedances should be equal. In the case of loudspeakers, a higher load impedance will give less power, but often less distortion too. Lower impedance will overload the amplifier output stage and could result in overheating and damage to the transistors.

Negative feedback If a fraction of the output of one or more stages is fed back out of phase to the input a gain reduction occurs. Additionally harmonic or frequency distortion generated within the feedback loop is reduced. As the fed-back signal is re-amplified and fed back again, the final result is the sum of an infinite series, but can be calculated from:

$$g = \frac{G \times n}{G + n} \text{ and } d = \frac{D \times n}{G + n}$$

in which g is the final gain; G is the gain without feedback; $1/n$ is the fraction of the output signal fed back; d is the final distortion; and D is the distortion without feedback.

Large amounts of negative feedback are used in modern amplifiers. Stages are designed with greatly increased gain to compensate for the loss attributable to feedback. Although rapid, feedback is not instantaneous; if a sudden transient signal occurs, it can cause overloading in the brief interval before feedback arrives to throttle back the gain of the controlled stage. The effect is termed *transient intermodulation distortion*.

The speaker generates back e.m.f. which appears at the amplifier output and so is fed with negative feedback to an earlier stage. It is delayed by cone inertia and contains distortions produced by the non-linear cone response as well as reactive elements in the cross-over network. As back e.m.f. is in opposite phase to the original signal, it becomes a positive feedback. Distortion is thus produced which is not evident when it is measured with a resistive load.

Some designs apply feedback in several loops over individual stages instead of a single loop over several stages. The trend now is to regard feedback as a necessary evil and to reduce it to a minimum.

Overloading   A stage is overloaded when the applied signal exceeds its designed input level and one half-cycle cuts the transistor off while the other drives it into saturation. It can be observed on an oscilloscope as clipping of the peaks. If one half-cycle clips before the other, this shows that the device is not biased to the mid-point of the characteristic. With output stages it indicates unbalance between the two transistors. Prior to overload the peaks encroach on to the upper and lower curves of the characteristic, rapidly increasing distortion.

Parameters   In the hybrid (*h*) parameters, small suffixes describe their function. The first suffix denotes the parameter, i.e.: F is the d.c. transfer ratio; *f*, the small signal transfer ratio; *i*, input impedance; *o*, output admittance; and *r*, the reverse voltage transfer ratio. The second suffix describes the mode: *b*, the common base; *c*, the common collector; and *e*, the common emitter.

The parameters for the common base and collector modes can be derived from the more frequently quoted ones for the common emitter by using the following formulae:

$$h_{fb} = \frac{h_{fe}}{1 + h_{fe}} \qquad\qquad h_{fc} = 1 + h_{fe}$$

$$h_{ib} = \frac{h_{ie}}{1 + h_{fe}} \qquad\qquad h_{ic} = h_{ie}$$

$$h_{ob} = \frac{h_{oe}}{1 + h_{fe}} \qquad\qquad h_{oc} = h_{oe}$$

$$h_{rb} = \frac{1 + h_{fe}}{h_{re} \times h_{ie}} - h_{re} \qquad h_{rc} = 1 - h_{re}$$

Resistor noise   Thermal noise is generated when current flows through any resistance whether it is composition, wire-wound, coil windings or the internal resistance of a transistor. It is dependent on temperature and resistance value. The formula is:

$$e_n = \sqrt{4kTR\Delta f}$$

in which e_n is average noise voltage; T is temperature in °K (°C + 273); k is Boltzmann's constant (1.38×10^{-23}); Δf is frequency bandwidth over which noise is of interest, for audio

it is 20 kHz; and *R* is resistance in ohms. Assuming a
temperature of 23°C, the noise voltage in µV becomes:

$$V = 0.0179\sqrt{R}$$

Threshold voltage   Below 0.65 V between base and emitter,
there is no controllable collector current for a silicon transistor.
Forward bias must exceed this, so a complementary pair must
have a difference between the bases of at least 1.3 V. The
threshold is very temperature conscious, so when operating
from a low-voltage source, operating point temperature-drift is
aggravated by that of the threshold voltage. It is thus better to
provide a large voltage drive in excess of the threshold via a
series resistor than a low signal voltage direct. Germanium
devices have a threshold of some 0.2 V and, as leakage current
drift is more significant, they are better operated from a low
impedance source to offset it.

Tone controls   Usually these provide continuously variable
treble and bass cut and boost. They can be passive, comprising
filter components in the signal path between two stages; or they
can be active, forming part of a frequency selective negative
feedback loop across a transistor or i.c. Logarithmic tracks are
used for volume, but linear for tone controls.

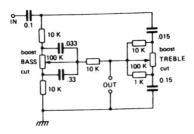

Figure 99. Passive tone control circuit.

Equalisers   Sets of controls that each boost or cut a single
different frequency are termed *graphic equalisers*, so called
because their purpose is to equalise anomalies in the frequency
response and the linear slider controls which graphically
display the settings. Commonly they have controls spaced at
one octave intervals, but these are too coarse to give anything
but approximate subjective equalisation, as the response dips

and peaks rarely fall exactly on the controlled frequencies. One-third octave spacing is required to give accurate compensation as the aural receptors in the human cochlea operate in one-third octave bands.

Each control has an associated tuned circuit which resonates at the required frequency. It can consist of a capacitor/inductor combination, but the inductors are bulky at the lower frequencies and are vulnerable to hum pickup. To avoid this, a circuit known as a *gyrator* is often used.

Gyrator   The gyrator consists of an inverting and non-inverting amplifier stage with the output of each connected to the input of the other. An impedance connected at one end produces an 'image' at the other equal to:

$$Z_2 = \frac{1}{Z_1 \times g_1 \times g_2}$$

in which Z_2 is the image impedance; Z_1 is the actual impedance; g_1 and g_2 are the slopes of the two amplifiers in amps/volts, which are also known as the *gyrator constants*.

If the actual impedance is a pure capacitance, of which the reactance is:

$$\frac{1}{2\pi fC}$$

then the image impedance becomes:

$$Z_2 = \frac{1}{\frac{1}{2\pi fC} \times g_1 \times g_2} \text{ or } = \frac{2\pi fC}{g_1 \times g_2}$$

This is the reciprocal of the actual capacitive reactance and the same as the expression for inductance: $2\pi fL$.

Therefore the image has the characteristics of a pure inductance of which the value in henries is equal to the value of the actual impedance in farads divided by the product of the gyration constants. It is worth noting from this that the image inductance is greater when the amplifier slopes are lower, but the Q is lower.

The image inductance can be combined with a real capacitor to form a resonant circuit. Provided the g of both amplifiers is the same, and the two capacitors have the same value, the resonant frequency of the circuit is:

$$f_r = \frac{g}{2\pi C}$$

and provided both amplifier input and output impedances are the same, with R being the value of the input impedance of one amplifier in parallel with the output impedance of the other, Q is given by:

$$Q = \frac{g \times R}{2}$$

An asymmetrical gyrator can be made from a single transistor as the inverting amplifier, and a resistor R_b across the collector to base as the 'non-inverting amplifier'. The Q is very low but the circuit is suitable as a supply-line series hum-filter. The image impedance is:

$$Z_2 = \frac{R_b}{g_1 \times Z_1}$$

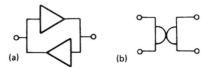

(a)   (b)

Figure 100. Basic gyrator: (a) block diagram; (b) symbol.

Transconductance   The conventional bi-polar emitter follower output stage produces output current from input current. Lower intermodulation distortion is obtained from common emitters or FETs. Output current is then produced from input voltage, hence the term transconductance.

Amplifier parameters

Output power   A 1 kHz input signal is increased until the output peaks start to clip or until a distortion level of 0.5 or 1.0% is reached with the specified load resistance connected. Both channels should be so driven for the measurement, otherwise the reading may be high owing to an abnormally high voltage from the partly loaded supply circuit.

The measured power which is usually quoted is the r.m.s. value. Peak power value is $\sqrt{2}^2$ or twice the r.m.s. and this has been used to increase the paper rating of low-power amplifiers. Another rating that has been used is music power, which relies on the ability of the power supply to momentarily provide higher voltages than normal due to the current stored in the reservoir capacitor. Thus, brief music peaks higher than the peak power rating can sometimes be handled without clipping.

The power rating holds good for only one impedance; at double the impedance it will probably be half, while at half impedance it will be more, though not usually double. Output power must be considered in conjunction with speaker sensitivity; a higher output being required for low sensitivity speakers.

Power bandwidth   This denotes the frequency range over which the power falls to no less than half (-3 dB) its maximum value. The range is less than the frequency response which is given at a much lower output level.

Damping factor   A speaker cone tends to oscillate for several excursions after the electrical signal has ceased before coming to rest. The coil generates a voltage that produces a current and a magnetic field which opposes the spurious motion and so dampens it. The magnitude of the current, hence dampening effect, depends on the internal resistance of the amplifier output stage through which it must flow. This resistance is much lower than the rated impedance, 0.1 Ω being typical. The damping factor is the speaker impedance divided by the amplifier source resistance. It ranges 25–1000, but the virtue of values over 50 is dubious because of the series resistance of the speaker speech coil and crossover network.

Frequency response   Usually measured at levels below 1 watt, this indicates the overall response and is greater than that at high power. To be meaningful the limiting levels should be given, although these are usually 3 dB. A response below 20 Hz is no advantage as rumble, record warp and pickup-arm resonance can be amplified and could damage the loudspeaker. Owing to transit time through the amplifier, negative feedback can become positive at supersonic frequencies. An induced r.f. or interference pulse can thus overload an amplifier stage severely when the h.f. response is excessive. A response much higher than 20 kHz can therefore be more harmful than beneficial.

Harmonic distortion   (THD) Any distortion of the signal waveform can be analysed as an addition of harmonics of certain order and magnitude. These are measured by injecting a pure, almost distortionless tone, notching it out at the amplifier output and measuring the residue which consists of noise and spurious harmonics. If the amount of noise is known, the distortion can be calculated from:

$$\frac{d}{S} = \sqrt{\left(\frac{r}{S}\right)^2 - \left(\frac{N}{S}\right)^2}$$

in which d/S is the distortion to signal ratio; r/S, the residue to signal ratio; and N/S, the noise to signal ratio.

Harmonics and music   The musical effect of harmonics differ. Some (second. fourth. eighth) are octaves and are perfectly harmonious. Others (third, sixth, twelfth) are at perfect fifth intervals from the fundamental and are not musically unpleasing if only the single note is played, but if the fundamental is the root of a diminished chord there will be a discord. The fifth harmonic is harmonious in a major chord but not in a minor, while the ninth is inharmonic. The seventh, tenth, eleventh and thirteenth all fall between Western musical scale intervals and so are unmusical.

Table of harmonics and their musical equivalents based on a fundamental of 220 Hz

Harmonic	Frequency	Note
	220 Hz	A
2nd	440 Hz	A^1
3rd	660 Hz	E^1
4th	880 Hz	A^2
5th	1100 Hz	C#
6th	1320 Hz	E^3
7th	1540 Hz	—
8th	1760 Hz	A^3
9th	1980 Hz	B^3
10th	2200 Hz	—
11th	2420 Hz	—
12th	2640 Hz	E^4
13th	2860 Hz	

Intermodulation distortion   When a high and a low frequency are amplified together by a non-linear amplifier, the low frequency modulates the amplitude of the high. The result is the generation of sum and difference frequencies, and further sum and difference frequencies arising from those, in a continuing series. These are not harmonically related to the originals, so the audible effect is worse than with harmonic distortion. It is measured by feeding a high and low frequency (commonly 100 Hz and 5 kHz, but others are also used) into the amplifier at a level ratio of 4:1. Output is filtered to remove the low frequency, and the high frequency is demodulated so leaving in any low frequency that had been modulated on it. This is measured and expressed as a percentage of the total signal. For a given amplifier, the amount is similar though usually greater than the specified harmonic distortion.

Sensitivity   This is the input signal that produces full output power, so a low figure indicates high sensitivity. Typical sensitivities are: gram, moving magnet, 13 mV; moving coil 0.1–0.2 mV; tape, 100–300 mV.

Signal/noise ratio   Noise including hum is related to a specific input. It is measured by injecting a 1 kHz signal with volume control at maximum and tone controls in the flat position. The signal level is adjusted to give maximum output and a voltage reading taken across a resistive output load. The signal is removed and the output meter switched down to read the residual noise with the input shorted, loaded or left open. The reading is expressed as a decibel ratio of the full power reading and is termed *unweighted*. When the residue is put through a filter that passes only frequencies that are subjectively annoying. it is said to be *weighted*. Typical values are: gram, moving magnet, 70–90 dB; moving coil, 65–85 dB; tape, 85–100 dB.

Slew rate   The response of an amplifier. especially its output stage, to a changing signal is not instantaneous. The slew rate describes the maximum voltage change that can occur in one μs. Typical values are 20–50 V/μs. It is most important for transient signals at high output levels.

Loudspeakers

The ideal loudspeaker would be a pulsating sphere that radiates sound in phase equally in all directions and at all audio frequencies. Such a device has not yet been achieved and all present designs fall short in some respect.

The driver

Moving coil The principle patented in 1898 by Oliver Lodge, later developed into a practical reproducer by Rice and Kellogg and patented in 1925, remains the principal type of speaker unit used today.

Conflicting requirements of the cone are *rigidity* and *low mass*. Rigidity is necessary for the cone to behave as a piston, moving the adjacent air without buckling and so adding spurious air motion. Low mass is required to minimise inertia and ensure the cone responds accurately to rapid signal changes. A large area is desirable to propagate low frequency sound, but this also conflicts with the low mass requirement.

Multiple drivers The latter problem is partly resolved by using two or more units of different size to handle different frequency ranges. The high unit, called the *tweeter*, is small and light and thereby able to respond quickly to the rapidly-changing high frequencies, while the bass unit, termed the *woofer*, is large enough to propagate the longer wavelengths. A mid-range unit of intermediate size, called a *squawker*, is sometimes used as well.

Doppler effect A high frequency produced by a cone that is also moving backwards and forwards with large bass tone excursions exhibits Doppler effect. The pitch rises as the cone moves forward and falls as it recedes, thus frequency-modulating it. The modulation is proportional to the amplitude of the low frequency excursions and, as small cones move further than large ones to produce the same power, they generate more f.m. distortion. The distortion increase is proportional to the square of the decrease in cone diameter and is at a maximum on axis, decreasing as the off-axis angle increases. It is minimised by having separate units for bass and treble, but still occurs with low and high frequencies within the pass band of each unit. Listening tests suggest that ±20 mm cone excursion is the lower limit below which Doppler f.m. distortion is inaudible. So it is likely to be noticeable only with insensitive speakers operating at high volume levels.

Phasing   To reproduce the original sound waves accurately. the phasing of individual units must be such as to preserve the original phase coherence when their outputs are acoustically combined. There are two problems: electrical and *spatial* phase differences. Reactive components in all but the simplest crossover network produce electrical phase shifts that change with frequency, resulting in phase differences between the drivers.

When mounted on a flat surface the bass-driver cone, being deeper, is further from the listener than that of the treble unit. The delayed sound results in a phase difference and, when the spacing is half a wavelength, cancellation. At normal spacings this can occur from 3–4 kHz, which is within the overlap region of the two drivers in many speakers. Some models have the tweeter connected in opposite phase to partly compensate; others have the units staggered, with the cabinet front in steps to bring all the units into line. These solutions are only effective on axis, as the delay reduces with an increase of off-axis angle.

Controlled flexure   Cone buckling, normally undesirable, is used and controlled in some cases to produce full-range speakers. The cone has curved sides, and the central area responds to high frequencies independently of the rest, owing to flexure at points governed by the cone contour, inertia of the outer area and the applied frequency. As the frequency decreases, larger areas are brought into play until the whole cone moves at the lowest frequencies. This effect enables a single unit to effectively reproduce a wide frequency range, 40 Hz–17 kHz being typical with some extending up to 20 kHz. Curvilinear contours are commonly used, but hyperbolic curves are thought to be superior and are also used. Flexure is encouraged by concentric corrugations in some cones.

Though covering less of the audible spectrum than multiple units, the full-range driver covers most of it, while phasing anomalies, distortion, ringing and interface problems caused by the crossover network, along with unit-sensitivity matching difficulties are avoided. Damping effect exerted by the amplifier is also greater.

Cone materials   The necessary characteristics are: rigidity, low mass and self-damping to minimise resonances. The most common is paper which is light and has excellent self-damping, but is not very rigid or consistent. The paper pulp stock consists of wood and rag with various additions. Among these are *kapok:* hollow, oily fibres from the silk-cotton tree chosen for their lightness: and waxes, fungicides and resins. The stock is beaten for precise periods in vats. Long periods produce short-fibre, thin, hard paper which results in sensitive cones, though prone to strong resonances. Short periods give long fibres that are more flexible and are suited for bass cones and controlled flexure full-range cones.

Bextrene is more consistent but has poor damping and needs coating with a plastic damper. Polystyrene reinforced with aluminium foil is very light and rigid but has poor damping, Honeycombed aluminium is about 1000 times more rigid than paper and has been used in disc form giving a more piston-like action than cones. Polypropylene is light, has good self-damping properties and is more rigid than paper.

Cone suspension   At the edge the cone is terminated by cloth, foam, rubber or simply by corrugations in the cone material. Its purpose is to support the cone, preventing sideways movement but without unduly restricting normal forward and backward motion. In addition it should absorb waves that travel out from the cone centre, which are otherwise reflected back and form standing waves. Cone material corrugations can be of *two sine rolls*, a *single sine roll*, or a deeper *accordion pleat*. Those made of foam or rubber can be of a *half-roll out*, having the roll facing the front of the speaker, or a *half-roll in*, with it facing the rear. Units designed for sealed-box mounting rely on the enclosed air to partly restrain cone movement and so are designated as having *acoustic suspension*. These should not be used in non-sealed enclosures.

The coil is centred within the concentric magnet pole pieces by a circular corrugated fabric restrainer connecting the back of the cone to the frame. Older models had a spider across the front of the coil which was secured by a centre screw to the centre pole piece, The coil could be centred by loosening the screw, inserting feelers inside the coil and re-tightening the screw. Modern units cannot be centred, but rarely go off-centre.

Cone resonance   Output drops at 12 dB/octave below cone resonance so it should be made as low as possible to achieve good bass response. The free air resonant frequency is proportional to the square root of the reciprocal of the mass of the cone and the compliance of the suspension. Compliance should be as large as possible commensurate with a stable suspension. Cone mass should be large and can readily be increased, but at a cost of reduced sensitivity and h.f. response. The latter is unimportant for a bass unit, but too low a sensitivity requires high amplifier power and produces high heat losses in the speaker coil. The resonant frequency can be obtained from:

$$f_r = \frac{1}{2\pi\sqrt{MC}}$$

in which M is the cone mass in grams; C is the compliance in m/N.

The air in a sealed enclosure adds to the cone stiffness so reducing the compliance and thereby raising the resonant frequency.

Compliance   The unit is the metre/Newton and is the reciprocal of the suspension stiffness: it can be calculated from the cone mass and free-air resonant frequency:

$$C = \frac{1}{(2\pi f_r)^2 M}$$

Damping   A peak appears at the resonant frequency which is Q times the normal level. Cone motion, hence the back e.m.f., is greater at this frequency, so a counteracting magnetic field is produced which is also greater. The peak is thus self-dampening, but the dampening depends on the effectiveness of the e.m.f. generation (flux density times coil length) and inversely on coil resistance, cone mass and resonant frequency. The value of Q is therefore:

$$Q = \frac{2\pi f_r MR}{(Bl)^2}$$

where R is the coil resistance; M is the mass in grams; f_r is the resonant frequency; B is the flux density; and l is the coil length.

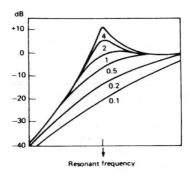

Figure 101. Damping of speaker resonant frequency with various values of Q.

A Q of 1 gives optimum damping at resonant frequency, but a lift of 3 dB just above it. A Q of 0.7 gives a flat response and an earlier but gentler roll-off and is the preferred value. As enclosure air-volume affects the resonant frequency it also affects the damping. There is thus a critical volume for a given bass unit to achieve the preferred damping level.

Cone velocity/radiation resistance As the frequency increases the cone inertia progressively reduces the amplitude of its excursions, so that at the high frequencies it is small compared to the low, for the same electrical input. However, at low frequencies the cone is an inefficient radiator, much of the air being pushed aside instead of being compressed. As the frequency rises the air does not move aside fast enough so it offers resistance to the cone and is compressed. The increasing radiation resistance exactly offsets the falling cone velocity so maintaining constant acoustic output at rising frequencies. The unit is said to be working in its *piston region* over this frequency range.

The radiation resistance rises until it is at a maximum at a certain frequency, but the cone velocity continues to decrease as the frequency rises further. Above this transition frequency there is a drop in output, but not immediately, as the beaming effect at higher frequencies concentrates the acoustic energy forward and cone flexure augments the h.f. output.

The transition frequency which ends the piston region for a flat disc radiator in a true infinite baffle is:

$$f = \frac{68,275}{\pi d}$$

in which d is the cone diameter in cm.

Delayed resonance Apart from controlled flexure, the vibrational modes of a flexible cone are quite complex. Ripples move out from the centre and, if not absorbed by the suspension at the rim, return. Standing waves occur producing resonances when the cone radius equals one wavelength or a multiple. For an 8 inch cone the first of such resonances is at 4 kHz. Energy is stored and subsequently released as spurious sound after the original cone excursion has ceased.

Bell mode In another motional mode, opposite quadrants of the cone flap, rising and falling together while adjacent quadrants flap in the opposite direction. However, two lines at right angles defining the edges of each quadrant remain stationary.

Dust cap/surround resonance The cap glued over the centre of the cone to exclude dust sometimes moves independently of the cone by reason of the compliance of the glued joint. It thereby has its own resonant frequency to colour the output. Some cones now have the cap moulded as an integral part to avoid this effect. Similarly the surround when glued to the cone can add its own coloration, but welding eliminates the problem.

Laser Doppler interferometer   Various methods of determining the cone vibrational modes exist. Lycopodium powder sprinkled over the cone produces ridge patterns. The stroboscope freezes the cone movement, or if the light flash frequency is slightly varied, it can be viewed in slow motion. Motion can only be detected over 1 mm, yet smaller movements occur and are audible. Laser holograms show movements comparable to the light wavelength by interference patterns, but cannot show larger ones: up to ten wavelengths is the maximum. Additionally, the direction of motion is not indicated. By combining a reference beam having an off-set frequency and a reflected one scanned across the cone, the difference arising from the Doppler effect is detected. This gives plus or minus values centred around the off-set, which thereby indicate motion direction. Three-dimensional plots can then be made from the result.

Speaker coil   Standard impedances, formerly 3 and 15 Ω, are now 4 and 8 Ω. The d.c. resistance is usually about two-thirds of the specified impedance. The impedance may be considered a minimum value, as it rises to a peak at resonant frequencies and is above the rating for most of its range. With multi-speaker systems the impedance may be mainly reactive at certain frequencies so that output current and voltage are out of phase. High current may thus flow at low power causing the amplifier current limiting circuits to clip prematurely.

Heat dissipation is a major factor; modern coils are wound on aluminium formers with high-temperature epoxy adhesives and can withstand up to 300°C. However, sensitivity can fall with temperature rise, as the resistance increases 0.4% per degree C. Thus the temperature of all units should be equalised to maintain tonal balance, but this depends on the programme spectrum as each dissipates different power levels. So although coils may be capable of withstanding it, heat should be removed as quickly as possible to prevent excessive resistance increase. Large magnet area helps in the long term: some coils are blackened and given heat fins to facilitate heat loss, but these measures have little effect in the short term.

Tweeter coils are most vulnerable because of their small coils and magnet assemblies. Maximum temperatures are around 120°C. To aid dissipation, some units have sealed gaps between the coil and pole pieces filled with colloidal ferromagnetic fluid.

The power shared between the treble and bass units differs according to the type of programme and also the distortion content, especially when the amplifier is clipping, as this increases the high frequency proportion. It is possible for a low-powered amplifier to heat a treble unit coil more than a high-powered one because it clips the peaks when operated at the same power level.

Some typical speaker coil temperatures with different programmes

	Temperature (°C)	
Programme	bass unit	treble unit
Piano *fff*	100	25
Orchestra *fff*	150	40
Heavy rock	120	75

Coil windings are usually of round copper wire, but aluminium is also used to reduce mass; ribbon wire wound edgeways-on achieves some 40% greater conductor density so improving motor efficiency.

Tweeter cones   Metal such as aluminium and beryllium has been used for its high ratio of rigidity to density which increases the sound velocity within the material. The frequency of the first break-up mode is thereby pushed higher. However, rigidity can produce a problem in that there is little or no flexure, so radiation resistance falls off earlier than with a less rigid cone, and the response suffers accordingly. Metal cones thus tend to be used over narrower bandwidths, being augmented by super-tweeters. Paper with a high kapok content and mica are the principal materials.

Domes   To avoid some of the problems inherent with cones and to give a wider angle of dispersion, domes are more in favour for tweeters.

Thiele–Small parameters   In the early 1970s Neville Thiele and Richard Small calculated and quantified the various parameters affecting the performance of infinite baffle and ported reflex loudspeakers, giving these a designation which has become a standard. In many cases the design of enclosures can be simplified by the use of these parameters. The main ones are as listed opposite.

Thiele–Small parameters

Bl	Product of flux density and length of coil in magnetic gap (not total coil length)	Tm
C_{ms}	Acoustic compliance of suspension	mN^{-1} x 10^{-6}
d	Effective piston diameter	mm
fs	Free air resonance	Hz
M_{ms}	Total moving mass of driver	gm
Q_{es}	Electrical Q factor	
Q_{ms}	Mechanical Q	
Q_{ts}	Total Q of driver	
Q_{tc}	Total Q of system including cabinet	
R_e	D.c. resistance of coil	Ω
S	Piston range sensitivity	dB
V_{as}	Volume of air having same acoustic compliance as the suspension	Litres
$\dfrac{(Bl)^2}{R_e}$	Motor factor	

If the total Q of a driver, Q_{ts}, is not given it can be calculated from the formula:

$$Q_{ts} = \frac{Q_{ms} \times Q_{es}}{Q_{ms} + Q_{es}} \text{ or } \frac{1}{Q_{ts}} = \frac{1}{Q_{ms}} + \frac{1}{Q_{es}}$$

The total Q designated Q_{ts}, is that of the driver only. The total including the cabinet is described as Q_{tc}. This of course cannot be specified by a manufacturer because he does not know the size of the cabinet in which the driver will be used, but it does appear in formulae which can be used to calculate the optimum cabinet size.

V_{as}, the volume of air having the same acoustic compliance as the mechanical suspension of the driver is a particularly useful parameter when quoted by manufacturer and is one of those that simplifies the calculation of enclosure volumes.

In the expression Bl, the flux density is in tesla. which is one weber per m^2. The density may be expressed in gauss which is one oersted (one line of flux) per cm^2. To convert, 1 tesla = 10,000 gauss. The length l, of the coil is in metres, but designations are commonly in mm to avoid very small fractions. Bl products range between 8–30 Tmm.

Enclosures

The front and rear waves generated by a loudspeaker cone are of opposite phase; compression at the front coincides with decompression at the back as the cone moves forwards. Unless kept separate, they would merge and cancel at the rim of the cone at all wavelengths longer than the radius of the cone, since that is the distance the waves must travel to meet. Frequencies below this point are attenuated at the rate of 6 dB/octave. The main purpose of the speaker enclosure is to keep the two waves apart for as long as possible, with the minimum modification to the front wave.

Baffle/doublet

Mounting a loudspeaker on a flat baffle increases the distance the front and rear waves must travel to meet and so extends the bass response. But to achieve a flat response down to 45 Hz requires a baffle with a radius of 25 ft (7.7 m), so full-range baffle speakers are quite impractical. A baffle with a radius of 2 ft (1.23 m), which is about the largest practical size for domestic use, starts to roll off at 280 Hz. However, the response is –6 dB at 140 Hz, and –12 dB at 70 Hz, which is a gentler slope than can be obtained from an enclosure. Wall mounting gives sufficient baffle area but can have other practical problems.

Because of the time taken by the sound to reach the edge of the baffle, delays occur which cause cancellation at some wavelengths and reinforcement at others. Cancellation occurs at wavelengths that are whole multiples of the radius, but reinforcement takes place at those that are 0.5, 1.5, 2.5, etc., times the radius.

The frequency response thus becomes a series of troughs and peaks when the radius from the cone to the baffle edge is the same in all directions such as when the speaker is mounted in the centre of a circular baffle. To avoid this the speaker should be mounted off-centre. Frequency irregularities from this cause are thereby smoothed out. Usually the driver is fitted in the upper part of a rectangular baffle. When so mounted, the wavelength at which the bass roll-off commences is equal to the shortest radius.

The advantage of a baffle is that the many resonances and colorations produced by even a well-designed enclosure are eliminated. A compromise is to fit shallow sides to form an open-backed box so extending the front-to-back path with the addition of minimum coloration. A further extension can be obtained by fitting a back having slots or holes which serve as an acoustic resistance, but at the cost of adding some coloration.

Polar propagation   of a baffle speaker is that of a doublet, which is a figure eight. Sound pressure at an angle from the front or rear axis is the product of that obtained on-axis and the cosine of the angle. At high frequencies the angle of propagation narrows. and the rear lobe is distorted by reflection and diffraction caused by the speaker frame and magnet. A flat baffle should be operated well clear of any wall otherwise the rear wave is reflected to reinforce and cancel the front wave at quarter and half wavelength spacings and their multiples, respectively.

Modification of the doublet propagation can be achieved by mounting two baffle speakers in a V configuration.

Infinite baffle

The rear wave is 'smothered' by mounting the speaker at the front of a sealed box. The enclosed air acts as a spring, reducing the suspension compliance and raising the resonant frequency. Mechanical suspension is made as compliant as possible to compensate partly, hence the term *acoustic suspension*. Compliance increases with box volume for a given driver of specified cone diameter and mass, to achieve a desired resonant frequency. The response falls at a rate of 12 dB/octave below resonance.

Sensitivity   of the infinite baffle is low because the mass of the bass speaker cone is generally made high to bring down the resonant frequency and so extend the bass response. Efficiency could be increased by making the magnet stronger, but this reduces the Q and results in overdamping, which in turn prevents the use of resonance to boost the bass. So bass roll-off begins at a higher point and the bass response is reduced. As with most of the other factors there is thus an optimum magnetic strength for a given set of parameters.

Air resonances   In addition to the main cone resonance, resonances exist that are functions of the three enclosure dimensions which should be dissimilar to avoid resonances reinforcing each other. These can be damped by absorbent material placed at the points of maximum vibration, termed the *antinodes*. For the fundamental this is halfway along each dimension. For the second harmonic it is at one-third and two-thirds the length, and for the third harmonic it is one-sixth, one-half and five-sixths. To damp all harmonics of all three dimensions thus requires virtually filling the inside of the enclosure with absorbent material. Material fixed to the inside of the enclosure panels has no effect on air resonances as these points are the nodes of minimum air vibration.

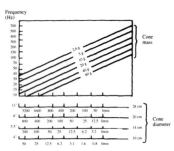

Figure 102. Enclosure volume in litres for given cone diameters and mass in grams for specific resonant frequencies.

Panel resonances   The high internal pressures in a sealed box cause the panels to vibrate at their resonant frequency and radiate sound. These vibrations can be minimised by a solid well-braced construction. High density panels of metal, brick or concrete have been used, as well as cavity panels filled with sand. They are usually damped by lining with a heavy absorbent. Some use light panels with heavy bituminous damping pads

Reflected wave   The rear pressure wave is reflected from the rear panel back to the speaker and out through the cone. The delay produces reinforcements at one-quarter and three-quarter wavelengths and cancellation at one-half and whole values, resulting in response peaks and troughs. An audible 'honk' can also be produced. Heavy damping of the back panel is the cure.

Damping   The degree of damping depends on the magnitude of the opposing force, which in turn depends on the efficiency of the system as a generator. Generator efficiency is governed by the magnetic flux density and the length of the coil, and inversely by the coil resistance, the cone inertia which depends on its mass, and the frequency of resonance.

The total Q comprising electrical Q (Q_{es}) and mechanical Q (Q_{ms}), which is designated Q_{ts}, is that of the driver only. The total including the cabinet is described as Q_{tc}. This of course cannot be specified by a manufacturer because he does not know the size of the cabinet in which the driver will be used, but it does appear in formulae which can be used to determine the optimum cabinet size.

If Q_{tc} is equal to unity, there is no peak at the resonant frequency because the amplitude of the cone excursion is just 1 times that at any other frequency. This would appear to be the ideal value. However, an undamped response consists of a peak that is sharp at its tip while being fairly broad at its base. If now we level the tip to unity value, there is still a slight rise on either side due to the 'foothills' of the base. The lower one disappears due to the bass roll-off below resonance, but the upper one remains.

A Q_{tc} of unity therefore produces a small rise just above the resonant frequency. The chart shows the effect (Figure 101). To eliminate this, we need a Q_{tc} that is actually less than unity. This is possible by reducing the cone diameter, or for a given driver, increasing the volume of air in the enclosure, which means increasing its size. The value usually chosen is 0.7 which gives a smooth roll-off, though one that starts above the resonant frequency.

Enclosure volume   The formula for designing a sealed enclosure for a particular driver is quite a complex one, but it can be greatly simplified if its use is restricted to a Q_{tc} of 0.7. For other values it is less accurate. The governing factors are: the compliance of the drive unit's suspension; the volume of air having the same compliance; the mechanical Q, Q_{ms}; and the electrical Q, Q_{es}.

To simplify matters further, these four factors are combined into two in the Thiele–Small parameters quoted in maker's specifications. The volume of a body of air having the same compliance as the suspension of the drive unit is given in litres and denoted by the term V_{as}. The mechanical and electrical Q are usually combined according to the previously given formula and designated as Q_{ts}.

So we need only the V_{as} and the Q_{ts}. The formula is:

$$V_b = \frac{V_{as} \times Q_{ts}^{\,2}}{Q_{tc}^{\,2} - Q_{ts}^{\,2}}$$

and as Q_{tc} is 0.7, then:

$$V_b = \frac{V_{as} \times Q_{ts}^{\,2}}{0.49 - Q_{ts}^{\,2}}$$

Not all bass drivers have the high compliance required for use in sealed enclosures, many are designed for open-backed systems. The formula is not intended for these and will give an incorrect answer if used for them.

Dimensional resonances   Being sealed boxes, infinite baffle enclosures are greatly affected by the internal air resonances. As shown earlier these are air resonances that are functions of the three enclosure dimensions. They occur at frequencies having half-wavelengths corresponding to the three cabinet dimensions.

If any two dimensions, or worse. all three are the same or nearly so, a very strong resonance is obtained. Hence height, width and depth must be different and not related so that one is a multiple or part multiple of any other. The golden ratio which applies to listening rooms (page 32), applies here too. So the ideal enclosure ratios are 1:1.6:2.5. or multiples of them.

Standing waves   When air pressure waves travel back and forth between two parallel surfaces that are half a wavelength apart, the areas of high and low pressure always appear in the same places and so it seems that the wave is stationary. Thus are produced what are termed *standing waves*. The antinodes or regions of maximum activity are at the halfway points while the lows or nodes are at the surfaces.

Harmonics have their antinodes at other points, so to suppress them all, absorbent material should be distributed throughout the enclosure. For small enclosures this is no problem, but with larger ones there can be compaction of the lower layers with the weight of the material. The effect can be minimised by standing rolls of BAF wadding vertically on end.

An irregular shape with non-parallel sides reduces dimensional resonance effects. The ultimate is perhaps the pyramid. At least one commercial loudspeaker was made in this form. The disadvantage is that a pyramid has only a third of the volume of a rectangle of the same base and height dimensions.

Compression distortion   As the cone of a bass driver moves inward from its rest position it first encounters little resistance from the enclosed air. Air is increasingly compressed as it travels so that resistance is correspondingly increased. The effect is like pushing in the handle of a pump with the exit hole blocked.

According to Newton's laws of motion, a body moves in proportion to all forces acting upon it. The increasing air resistance which opposes the applied electromagnetic force impelling the cone, must therefore progressively reduce the distance it travels for a given input.

The response to the input signal is thus not linear and the tops of applied waves are rounded off. The same effect occurs when the cone moves outward, but then the cone is restrained by an increasing vacuum. Distortion is the obvious result.

With short wavelengths the cone excursion is complete before the compression wave has filled the enclosure, so the compression has little effect on the result. It is the longer wavelengths corresponding to the bass frequencies that give rise to the effect. Distortion plots for IB enclosures typically show a rapid rise of 4 or 5 times the average THD below 100 Hz.

Reflex enclosure

Here the enclosure has a vent or port at the front through which the rear sound can escape. Usually the port has a short inlet pipe. The air in the pipe can be considered as a wad having mass, inertia, and its own resonant frequency. The frequency is adjusted to be the same as that of the cone/enclosure air volume resonance; so there are two resonances tuned to the same frequency. The effect is similar to that of the Helmholtz resonator which resonates at a frequency determined by the volume of the cavity and the length of the neck.

At high frequencies the inertia of the air wad is too great for it to respond, so the enclosure behaves as a totally sealed unit. At resonance the wad reacts against the springiness of the internal air and vibrates, but does so in phase with the cone excursions. The output thus reinforces that from the cone. Cone excursions are actually reduced at resonance thereby increasing the power handling capability.

Below the resonant frequency, the air wad has insufficient inertia to restrict the pressure waves which emerge out of phase and so cancel the front waves. Roll-off below resonance is therefore rapid.

The formula (in Imperial measurements) for calculating the volume of a reflex enclosure is:

$$V = \pi r^2 \left(\frac{4.66 \times 10^6}{f^2(L+1.7r)} \right) + L \quad \text{in}^3$$

in which V is the enclosure volume; L is the length of the pipe (in inches); r is the effective cone radius (in square inches); f is the cone resonant frequency. In this case the area of the port must be the same as that of the cone.

The formula for metric measurements, with all dimensions in cm, is:

$$V = \pi r^2 \left(\frac{304 \times 10^6}{f^2(L+1.7r)} \right) + L \quad cm^3$$

A simpler alternative formula for calculating the volume of the enclosure in litres (a litre is 1,000,000 cubic millimetres or 61 cubic inches), using Thiele–Small parameters is:

$$V_b = 20V_{as} \times Q_{ts}{}^{3.3}$$

in which V_{as} is the volume of air having the same compliance as the drive unit suspension, and Q_{ts} is the total driver Q. The fractional power can be obtained with a calculator having a x^y key.

This does not give the tube area and length, and as the two resonant systems must be matched, the dimensions are critical. The resonant frequency depends in the mass of air in the tube, and also its compliance. If mass was the only factor, the calculation would simply involve the volume of air in the tube, which is proportional to its area and length.

The complication is increased by the compliance which is the inverse of the resistance offered to the rush of air trying to get through it. A wide, short tube offers less resistance, and so has a higher compliance, than a long, narrow one, yet the mass of air could be the same in each.

A further factor is that a 'stub' of air develops beyond the end of the air in the pipe which moves it, so forming an invisible extension, increasing its mass. This effect varies with the length of the tube.

Unlike the totally sealed box in which size has little effect on the bass response as defined by the f_3 point (the point at which the output falls to –3 dB), increasing the size of a reflex cabinet within limits, does give a lower f_3, and for a given diameter, requires shorter tubes.

It may seem that a *longer* tube containing more air would be needed to balance the larger mass of air in the larger enclosure, but the greater air mass is more easily compressed and so has a higher compliance. In the case of a tube open at both ends, a short tube offers less resistance to a volume of air flowing through it than a long one, and so has a higher compliance. Thus the compliance of the tube is increased when the tube is shortened to match the compliance increase with larger cabinet size.

The tube length can best be determined experimentally by trying different lengths of 3 inch drain pipe while sweeping the bass frequencies with an audio oscillator and measuring the impedance with a series ammeter to give twin dips.

Impedance The impedance of a bass driver rises to a peak at resonance because the more vigorous cone movement generates a larger back e.m.f. which opposes the signal current. Current is therefore less for the same applied voltage which by Ohm's law is equivalent to a higher impedance. With reflex operation, the cone movement is restricted at resonance, so a dip appears in the impedance curve instead of a peak. On either side of resonance a small rise corresponding to the side slopes of a resonance peak produces two small peaks. The effect may be considered as a single peak with the tip inverted due to the reflex action, which is effective only over a very narrow band centred around the resonant frequency.

They appear as such only when the two resonant systems are at the same frequency which is when the reflex action is working properly. With small enclosures, some designers tune the tube air mass to a slightly higher frequency than that of the enclosure, and this shows up as a pair of unequal peaks, the lower being the largest.

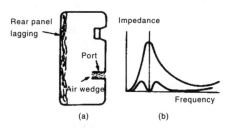

Figure 103. (a) reflex enclosure; (b) impedance peak of sealed box and reflex enclosure, the latter having two spaced peaks either side of resonance.

Auxiliary bass radiators The ABR is a loudspeaker cone in a frame without the coil and magnet. It is fitted in place of the port and pipe, and the drone cone takes the place of the wad of air. Otherwise the action is the same, the cone moving in phase with the speaker cone at the resonant frequency. The mass can be made large compared to an air wad, so a low resonant frequency can be achieved with a smaller enclosure.

The horn

Air impedance is low whereas that of the cone is high, so coupling is inefficient. Matching is greatly improved if the cone/air interface is via a duct of exponentially increasing area. This gives optimum matching and prevents internal reflections.

The shortest wavelength reproduced is twice the diameter of the throat, which must be small to achieve a good high frequency response. The longest wavelength it will radiate is twice the flare diameter. So a large mouth is required to obtain good bass. For a response down to 100 Hz, a flare of 5.6 ft (1.7 m) is needed.

As the area of the horn increases exponentially to the flare, length is related to the area of the flare. It is thus also related to the lowest frequency produced which is also the resonant frequency of the air in the horn. Required length is given by:

$$L = \frac{\log A - \log a \times 4000}{f \times \log \varepsilon}$$

in which L is the length (cm); A is the area of the flare (cm^2); a is the area of the throat (cm^2); f is the lowest frequency; $\log \varepsilon$ is 0.4343.

Hyperbolic flares   Hyperbolic expansion gives a response down to a lower frequency than the exponential, but the fall-off below it is more rapid. The expansion from the throat is more gradual which builds up pressure there and gives more distortion.

Throat configuration   The area in front of the cone narrows in order to provide a small throat area and achieve a good high frequency response. A region of high pressure is thereby created which is balanced by a sealed chamber behind the cone to equalise the pressure on it and so reduce non-linear distortion. A plug with holes is often introduced in front of the cone to delay pressures from different parts of the cone equally as they enter the throat, thus preventing cancellations.

Efficiency   A principal feature of the horn is efficiencies up to 80% so requiring little amplifier power. It is also highly directional.

Public address   The large full-range metal horn has its principal application in public address work. The length, which makes the unit unwieldy, is reduced in the re-entrant horn by coiling it up, but it remains difficult to manage because of the large flare required to obtain a good l.f. response. Usually units of more practical size are used at the expense of the lower register, which results in many p.a. systems sounding hard and unpleasant. For quality installations, they have been superseded by the line source column.

The horn can be used to advantage for reproducing high frequencies only, and is so employed with the ribbon tweeter. The output from the ribbon is low, so the high efficiency of the horn gives the required sound level.

Folded horn Some domestic hi-fi systems use the horn principle. These are often rear-loaded having a folded passage of increasing area formed behind the driver by wooden partitions in various ingenious configurations. When operated in a corner, the room walls and floor serve as the final part of the flare so overcoming the size problem. The front of the driver radiates the h.f. directly. Such designs are compromises as progression is only roughly exponential and the length is short, so internal reflections and colorations can occur. Even so, efficiencies of some 40% are obtained compared to around 0.5% for a typical IB enclosure.

Labyrinth/transmission line

One of the most satisfactory solutions to the rear wave disposal problem. The enclosure contains a long passage formed by internal baffles similar to the folded horn, but the passage is longer and the area does not increase. The operation is similar to a long, 'lossy' electrical transmission line in which most of the energy is lost and very little appears at the end to be reflected back.

Length of the line is made a quarter wavelength of the resonant frequency of the bass driver, or a little longer. The sound appearing at the end then is in phase with the front wave and reinforces it as a reflex enclosure does. Ideally, the area of the line should be the same throughout its length, being equal to the area of the cone. Inasmuch as the line is folded, resonances are likely over the individual sections due to back reflections, as well as the complete length, especially if the bends are sharp. These can be spread by tapering the area along its length, but can better be dealt with by arranging for the sound to be completely reflected around the bends.

The bends are often left without means of reflection, but the performance is improved by reflectors placed at a suitable angle in the bend. These usually consist of a flat wooden board, but being in the direct path of the pressure wave, they can absorb part of the energy and set up vibrations at their own resonant frequencies, thereby adding coloration. Ceramic tiles have proved very successful for reflection especially if bedded on a shallow concrete wedge in the corner of the bend. For a U bend, two tiles in a V formation at a 90° angle to each other give excellent results.

Resonant pipe   The transmission line also behaves as a tuned resonant pipe closed at one end. Such a pipe has a strong fundamental resonant frequency which is utilised to produce an in-phase output at the vent. However, any resonant chamber stores energy which is released after the source has ceased. To reduce coloration from this cause, the pipe should be well damped by filling the pipe with absorbent material. The node is at the closed (loudspeaker) end and the antinode at the open end. By packing the material more densely at the vent, damping is effected.

A closed pipe generates no even harmonics, so there is no second. The third harmonic has antinodes at the one third and open end positions. The packing arranged at the vent for the fundamental also serves to dampen the latter, and an increase in density at the one third position takes care of the first antinode. There is no fourth harmonic and the fifth is sufficiently weak to be well damped by the general filling There is no need to lag the panels or divisions because there are no high pressure differences across them. Panel resonances are not a problem.

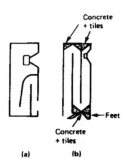

Figure 104. (a) Radford transmission line; (b) Kapellmeister speaker.

Column/line source

These speakers are used for public address purposes and consist of a vertical column of drivers in a narrow enclosure.

Polar propagation pattern A single driver in a cabinet has a roughly conical front propagation pattern. When arranged in a vertical column, these successively overlap to give strong reinforcement at the centre with minimum reinforcement at the ends. The result is a slightly divergent beam with a flat top and bottom, and a wide horizontal angle. There are small vertical lobes close to the column owing to the end units. It can thus direct sound into an audience with the minimum radiated to the upper walls and ceiling to give unwanted reflections.

Power tapering The end lobes can be eliminated by tapering the power toward both ends of the column. In practice, the end lobes are not usually detrimental: in fact the lower one can be an advantage as it gives coverage below the plane of the column in the near field where sound pressure is low. Thus it serves the first few rows of the audience.

Frequency response and polar propagation The actual interference pattern is a complex one that is frequency dependent. In addition to the reinforcement at the beam centre, there is cancellation at its top and bottom at 29° from the axis, when the spacing between the end units equals a wavelength. That gives a total beam spread in the vertical plane of 58°. At double the wavelength the spread is 29° and at four times it is 14.5°. So at very short wavelengths compared to the column length, the vertical angle narrows to a pencil beam. Only a restricted audience area that is on axis from the centre of the column will thereby receive high frequencies. For example a 5 ft (1.5 m) column has a wavelength of 220 Hz at which the vertical dispersion is 58° and at 440 Hz it is 29°.

For speech the high frequency dispersion of a 5 ft column is generally adequate, but with long columns sometimes used in large auditoria this may not be so, especially as it is difficult to arrange for *any* part of the audience to be on axis. In such cases *frequency tapering* may be necessary. Coils are wired in series with the end few units, so restricting the high frequencies to the column centre. The column thus becomes short for the high frequencies and long for the low.

Attenuation with distance At a single point in the near field, the sound pressure is from one or two units. As the distance increases so does the pressure, as it encounters the overlapping fields from adjacent drivers until it reaches a maximum when it is in the area of total reinforcement from all units. The distance of this point from the column is proportional to the length of the column. From there, the sound pressure is attenuated at a rate of 3 dB per doubling of distance. This is in comparison with the 6 dB of a single unit.

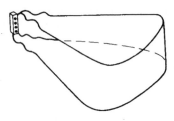

Figure 105. Nominal propagation pattern for line-source speaker. Actual vertical dispersion varies with frequency.

Loudspeaker absorbents

Enclosures that have high pressures built up inside such as the infinite baffle and bass reflex need absorbent material fitted to the panels to dampen panel resonances. Rear panels need more damping to suppress the reflected wave which otherwise travel back to, and out through, the cone of the bass driver. Generally, the thicker the layer, the lower the frequency it will effectively absorb. Layers of insufficient thickness are of little use as it is mainly the mid- and lower-frequencies that cause panel resonances.

Air resonances also occur related to the three enclosure dimensions in infinite baffle boxes. For these, damping is required at the antinodes, the regions of maximum air particle velocity, which are in the central areas of the enclosure, Air resonances are not affected by material fixed to the panels as these are at the nodes of fundamentals and harmonics. A large part of the internal volume is thus taken up with damping material. Transmission line speakers also need to be filled with absorbent to dampen the fundamental and odd harmonics, but do not need panel lagging.

In the case of the reflex enclosure, padding is needed on the rear panel and is sometimes fixed to the others, but no space filling, as the design relies on air resonance for the reflex effect.

Adiabatic propagation   Pressure produces heat, so sound pressure waves have higher temperatures in the areas of high pressure than in the troughs of low pressure. The areas at higher temperature expand and increase the pressure, producing more heat. Sound velocity increases with temperature rise, so it is pushed higher by its own pressure. Eventually, the adjacent areas of temperature difference merge and equalise, but not fast enough to affect the velocity of the sound wave which has passed on. The propagation is said to be *adiabatic*, which is impervious to heat change.

Isothermal propagation   If sound travels through a medium that conducts heat more readily than air, the temperature differences are quickly merged, the pressure falls and the velocity drops. This condition is described as *isothermal*. Most of the materials used for sound absorption, although considered poor heat conductors, conduct heat better than air. Thus partly isothermal conditions exist in enclosures filled with these materials.

The reduction of sound velocity has the same effect as having a larger cabinet or a longer transmission line. Bass resonance is thereby reduced. A fully isothermal state would produce a sound velocity reduction of $\sqrt{2}$ or 1.414. This would reduce the resonant frequency to about 0.833 of its former value. However, because waddings are not good conductors the effect is only partial, but it is a useful bonus.

Materials   Polyurethane is commonly used because of its convenience, but bonded cellulose acetate fibre (BAF) is probably better and is also supplied in thick sheets. Glass fibre is another material, but is inferior to the others, and there may be a health hazard in working with it. Long-fibre wool is considered to be the best of all, but being loose is not so easy to use, and needs moth-proofing. BAF sheeting is usually supplied in 1 inch thickness, and it can be rolled up to form large wads, or laid out in several thicknesses. It is effective and convenient.

Electrostatic speakers

The electrostatic speaker uses a thin conductive diaphragm driven over its whole surface by an electrostatic force. As rigidity is not a requirement the mass can be extremely small, so the response to transients and high frequencies is almost immediate.

With all parts of the surface being driven, there are no break-ups or spurious movements such as experienced with a cone. In Britain, QUAD is virtually synonymous with the electrostatic speaker, but there are many other makers in the field, especially in America.

Bias   When a voltage is applied across two conductors having a large surface area in close proximity they are mutually attracted. If one is flexible it will move toward the other. An applied a.c. wave causes one such movement for each half-cycle hence two per cycle, thereby producing frequency doubling. This is avoided by superimposing the a.c. on a fixed d.c. bias so that it adds and subtracts from the steady voltage on alternate half-cycles.

Push-pull   At bass frequencies the flexible member must make large excursions which considerably changes the spacing. As the electrostatic force acting on it varies with the spacing, the force is not constant but changes with the amplitude of the applied signal. Thus amplitude distortion is generated.

By placing the flexible diaphragm between two rigid members that are acoustically transparent, a push-pull device is created. Signals of opposite polarity are applied with the bias between the diaphragm and each of the rigid members.

At each half-cycle, attraction, hence distance, to the one plate is increased while that to the other is reduced. The diminished force at one side is thus balanced by the increased force at the other.

Charge migration   The attraction depends on the applied voltage and to the electrical charge on the diaphragm. If the latter is constant, then only the applied voltage influences the movement. However, as the distances between the diaphragm and the fixed members change, so does the capacitance and the charge flows in and out accordingly. The varying charge thus modifies the attraction and generates distortion. The effect is eliminated by including a series high-value resistor in the bias supply. The time constant of the RC combination is then very long compared to the lowest frequency (minutes), so very little of the charge flows and the diaphragm movement becomes proportional only to the applied voltage.

The charge can become localised, flowing from some parts of the diaphragm to others nearest an outer plate. Attraction can thus vary over its surface, resulting in unequal movement. To prevent this, the conductive coating in the plastic membrane is made highly resistive, having one conductive atom to ten million non-conductive ones, with a resistance of hundreds of megohms.

Polarising voltage   A high polarising voltage of several kilovolts is required (5.25 kV with the Quad) and the signal voltage can be of similar order, being obtained from the amplifier via large step-up transformers. There is thus the possibility of destructive flash-over. Reducing the voltages to a safe level limits sensitivity and output. In one model the unit is enclosed in inert gas which inhibits flash-overs and permits higher voltages and greater output power.

Another method of protection used by Quad, is to sense ionisation, which precedes a flashover, by an aerial running around the frame. When detected, a triac crowbar circuit is immediately activated, which short-circuits the amplifier output with a 1.5 Ω resistor.

There is also a limiter that reduces the applied signal voltage and adds distortion as an audible indication when the input exceeds 40 V into a nominal 8 Ω. A further protection against flashover is to use thermoplastic outer members that are coated with conductive material on the outside.

Bass response   All electrostatics are doublets, radiating from both sides like a moving coil speaker on a flat baffle. They cannot be housed in an enclosure as the internal pressures would have an even greater effect on the diaphragm than they have on the cone of a moving-coil unit.

The doublet propagation pattern restricts the low frequency response because of front and rear wave cancellation effects. To obtain a good bass response full-range units must therefore be large.

Room resonances   A conventional speaker in an enclosure has a roughly omnidirectional propagation pattern at low frequencies. It thereby excites all three room resonances corresponding to height, width and length. This often increases the apparent bass response if one of the dimensions has a fundamental or harmonic wavelength in the lower register of the speaker, although the effect is unnatural. The doublet has a figure-eight propagation pattern and most of the energy is radiated as particle velocity along the front and rear axes rather than pressure waves in all directions.

It thus excites only one of the room resonances and even this can be minimised by angling the speaker. The effect can sound deficient in bass, adding to the already reduced bass due to side cancellation, but is far more natural and free from coloration.

High frequency response   The ultra-lightweight diaphragm can produce high frequencies in the supersonic range. Sound waves propagated from parts of the diaphragm that are a half wavelength difference in distance cancel each other. With a large width, many such differences appear as the listener moves off axis and high frequencies having wavelengths of this order are only coherent on axis. So a narrow forward h.f. beam results.

In the original Quad electrostatic this was overcome by using three separate units: a narrow centre one handling the high frequencies and two outer ones reproducing the bass. This required the use of a crossover network, one of the things which a full-range speaker seeks to avoid.

In the latest models, the conductive diaphragm coating is divided into a series of five concentric rings with a centre and two outside segments, making eight in all. These are fed from a sequential delay line so that the signal is applied first to the centre and then consecutively at 24 μs delays, with attenuation, to the other sections.

A spherical-fronted wave is thereby produced which simulates a point source situated 30 cm behind the diaphragm. Such a source has a wide h.f. dispersion giving excellent off-axis response as well as good stereo imaging.

Loudspeakers for electronic instruments

Drivers Musical instrument loudspeakers are quite different from hi-fi models. The first are used for the *production* of sound, the second for *reproduction*. In the latter case, all forms of distortion must be kept to a minimum and the response to all frequencies must be equal. The loudspeaker is neutral, contributing nothing and omitting nothing.

With electronic instruments there is no original sound to reproduce, the loudspeaker is part of the instrument. As the sounding box of an acoustic guitar adds overtones and resonances giving it character and timbre, so does the loudspeaker with the electronic instrument.

Loudspeakers for the lead and rhythm guitar in particular, are deliberately made with in-built distortion, and cabinets are often constructed so that drivers can be easily exchanged to give a different sound. One of the methods used is to employ paper cone surrounds instead of cloth or rubber, to encourage reflections and standing waves.

Range Musical instrument loudspeakers need cover only the range of the instrument with some treble extension to produce overtones and harmonics. Often, a single driver is all that is needed. Hi-fi units must cover the whole range of audibility from deepest bass to highest treble overtone. Full-range loudspeakers are required for wide-range keyboard instruments and stage monitors.

Cone size Cone sizes range from 10 inch to 18 inch and with the 10–12 inch ones, up to four units are often mounted in the same cabinet. Power output is much greater than with domestic units, so the drivers must be capable of handling it.

Tweeters With full-range units there is a problem in obtaining a wide dispersion of high frequencies to cover all parts of the auditorium. A tweeter diaphragm propagates high frequencies in a narrow conical beam with about a 60° dispersion angle. A rectangular horn fitted to the front extends the side angle to some 90° at the expense of the vertical which is reduced to around 40° depending on the dimensions of the horn.

Another type is the bullet tweeter which as its name implies has a bullet-shaped diffuser surrounding a circular horn in front of the diaphragm. This radiates a conical beam over a 90°

angle. A wider dispersion can be obtained with a slot diffractor which has a horizontal spread of over 120° and a vertical one of 50°.

As the sound energy is thus spread, the tweeter has less 'throw' than when the sound is concentrated into a narrower beam.

High frequencies are attenuated with distance to a greater extent than low frequencies, they also undergo greater absorption by the clothing of the audience. So, the sound reaching the back rows will be deficient in the highest harmonics. One remedy for large auditoria is to have an array of tweeters in an arc, mounted in a separate box from the bass unit. This gives a wide dispersion while retaining the long throw of individual units. The spacing and angling of the units can be critical if mutual interference is to be avoided. Wide vertical dispersion is unnecessary unless a balcony is to be served from the same rig.

Mid-range   Bass units of 15–18 inches do not perform very well in the middle frequencies, so mid-range units are usually needed. However, deep bass can be obtained from 12 inch rubber edged drivers which also have a good mid response. These need more power for the same output than cloth or paper edged surrounds.

The use of mid-range units avoids Doppler effect in the bass unit at higher powers. This can become noticeable when bass cone excursions are large and are accompanied by frequencies in the higher mid-range. At more moderate powers the amplitude of the bass cone is insufficient for the Doppler effect to be detectable.

Lead guitar loudspeaker   The frequency range is from 196 Hz (G_1) to 1,568 Hz (G^3), so a deep bass response is not needed. A reflex enclosure is unsuitable, and an infinite baffle unnecessary, although it can be used. An open-backed unit, that is one having small holes in the back to serve as an acoustic resistance, is very suitable.

It should be noted that an open-backed cabinet can be fitted with an acoustic suspension loudspeaker only if it is never used for bass. Hence the combination can be used for the lead guitar.

The driver should be a 10 or 12 inch unit, with in-built distortion and non-linearities to add the timbre. Units with a paper surround, that is the cone pleated at the edge and joined directly to the frame, provide these. A range of such drivers are available for the lead guitar, each having different distortion characteristics. There are also cloth surround units for those who prefer a smoother tone.

Bass guitar loudspeaker   Frequency range is 41 Hz (E_3) to 261 Hz (middle C), so loudspeakers producing the deepest bass are necessary, with very little treble if any. Ideally, the loudspeaker should go down to 41 Hz, the lowest note.

Many driver units for use with bass instruments do not extend down to 40 Hz, but tail off around 60 Hz, half an octave higher. Their response though does not end abruptly at the specified frequency, it rolls off below resonance at 12 dB per octave for infinite baffle enclosures and 24 dB per octave for reflex.

The fundamental frequency of most vibrating strings is accompanied by large amounts of second and third harmonics with diminishing amounts of higher ones. The fundamental thus provides only a small part of the total energy of the note, and if cut out altogether is hardly missed.

So, bass loudspeakers with a response to only 60 Hz or thereabouts perform satisfactorily. However, for maximum effect at deep bass notes, the response should go down to 40 Hz. This is best achieved by means of a reflex loudspeaker. If two or more bass loudspeaker cabinets are stacked, the bass response goes lower. Two will extend it by about 10 Hz, and four by 20 Hz.

For a full deep bass with full production of the fundamentals 15 or 18 inch drivers are needed. However some performers prefer using 12 or even 10 inch drivers to get a faster response with more attack, fitting four to a single reflex cabinet to get a good bass response.

The bass instrument does not need a distinctive sound, so does not require in-built distortion. Players usually prefer the sound to be smooth and with a uniform response over its range. To achieve this the driver cone should have a cloth surround rather than paper.

Some bass players like a brighter tone, and for these 15 and 18-inch drivers with paper edges are available. Other models have rubber surrounds which give even less coloration than cloth and deeper bass. However they are much less sensitive, needing around three times the power of the cloth and paper types for the same sound output. As there is no need for the bass response to go below the lowest note of the instrument, the low sensitivity reduces their appeal for this purpose.

Keyboard   An electronic concert or church organ has pedal notes going down to C_4 which has a frequency of 16.3 Hz, but most keyboard instruments have a more modest bass range. The treble range of most keyboard instruments extends an octave or more higher than that of the lead guitar, and many of their stops give effects that are rich in high order harmonics. The loudspeaker response must therefore range from the lowest bass note to the highest harmonic.

Auditorium size is another factor to be considered. Large ones need a stack of separate bass, middle and treble units whereas smaller halls can be served by one or two full-range loudspeakers.

As all the tone colours have been achieved electronically, no extra coloration is required from the loudspeakers, even though they are part of the instrument. So the loudspeakers should be up to hi-fi standards, using cloth or rubber surround bass drivers.

Monitor When a group are playing behind a stack, it can be difficult to hear just what the effect is, especially if the audience is noisy. A monitor or 'fold back' loudspeaker is thus necessary to feed sound on stage. This needs to be a full-range unit although extreme bass and treble are not necessary. High power is not essential, and could lead to feedback with the on-stage microphones if the volume is too high.

An infinite baffle, or reflex can be used. A two-way system consisting of bass unit and tweeter is suitable, but a twin-cone full-range driver is also satisfactory.

A wedge shape is much favoured for the cabinet with the front sloping upward. It can thus be placed on the floor of the stage, and be directed toward the heads of the performers. A useful addition is a volume control which enables the level to be adjusted during performance without upsetting the balance of other units that may be operating from the same amplifier. Special high power controls of 30 watts rating are obtainable for the purpose.

Vocal Although not electronic instruments, vocals are an integral part of group performances and so are here considered. The human singing voice ranges from about 73 Hz, the lowest note of a bass, to 1,046, the top C of the soprano. However, pop and cabaret singers have a much more restricted range than that. Harmonics up to around 5–6 kHz need to be reproduced to give intelligibility, and preserve the character of the voice.

A loudspeaker for vocal use thus does not need to have an extended frequency response. It may be noted that female voices. although having higher-pitched fundamentals, have fewer harmonics than the male, and so do not extend as far up the frequency spectrum. This limited harmonic content is why female voices usually sound softer, whereas the male with a larger number of harmonics sounds harder and more incisive.

It is also the reason why male voices are more understandable to the hard of hearing, not because they are louder, but because they have more harmonics that give greater clarity.

The reproduction needs to be uncoloured and without distortion. Reverberation may well be added at the amplifier, but the loudspeaker itself should reproduce no more or less than what goes into it.

The open-backed cabinet suggested for the lead guitar is very suitable here as it gives good, low-distortion reproduction lacking only in bass, which for vocals is not needed. Full-range infinite baffle loudspeakers are sometimes used, although they are not really necessary.

A paper-edged lead guitar driver should not be used because of the in-built distortions. It should have a cloth or rubber surround.

Cabaret   Cabaret turns often have to perform in restricted venues and to smaller audiences than groups, so the equipment should reflect this. Loudspeakers can be smaller and less bulky to transport. If the turn is only vocal, a single 10 inch full-range driver in an open backed cabinet, should serve the purpose well. Two cabinets are usually required to ensure the area is fully covered, but may not always be required. High power is not usually necessary.

If the stage is low or non-existent, loudspeaker cabinets placed on the floor will be masked by the front of the audience. As there may be no suitable objects on which to stand them, an essential item of equipment is a loudspeaker stand for each unit. These are like a more substantial version of a microphone stand with fittings at the top to secure the loudspeaker, and are collapsible for transporting.

A loudspeaker on a stand does not excite bass resonances in the platform which can happen when a loudspeaker unit is standing directly upon it, and so does not produce a bassy, chesty sound, but is clearer and more natural.

Feedback between a floor-standing loudspeaker and microphone can take place through the flooring. This is less likely when the loudspeaker is mounted on a stand, especially if the stand has rubber feet. Feedback can and still does occur through air coupling, but the problem is reduced when one possible path is thus eliminated.

If the turn is instrumental, the loudspeaker should be chosen to reproduce the range of the instrument. Higher power than is needed for a purely vocal sound is required because the sounds made by many instruments have large starting transients which could produce distortion in an under-powered system.

Unconventional speakers

Orthodynamic drive   This has a diaphragm driver similar to an electrostatic, but impelled by a magnetic field instead of electrostatic force. It thus has the advantage of an electrostatic without the need for high polarising voltage and risk of flash-overs. The plastic diaphragm has a pattern of copper or aluminium conductors etched by printed circuit techniques. The magnetic field is produced either by short individual bar magnets, or ceramic magnetic material plates with holes, and zones magnetised between them. The magnets or plates are mounted either side of the diaphragm with the poles facing it, alternate poles being of opposite polarity. The rear poles are connected by a magnetic link so completing the magnetic circuits.

The poles on the opposite side of the diaphragm are of the same polarity. By this arrangement, the fields are confined to the space around the diaphragm and interact with the fields generated by the current flowing through its conductors. The conductor pattern cannot be of a simple spiral or zig-zag because current in adjacent tracks would be going in the opposite direction and would produce opposing magnetic forces. Loops that double back and similar devices must be used to ensure that all conductors within the same field direction carry current in the same direction.

Orthodynamic tweeters which have a long vertical diaphragm have been produced to give maximum h.f. side dispersion on which the conductor is etched as a concentric coil. The principle has been used in headphones and in mid-range units.

Heil air motion transformer   The drive uses the orthodynamic principle with an etched conductor on a plastic diaphragm operating within applied magnetic fields. There are two differences, the fields are at right angles to the diaphragm instead of being in the same plane and the diaphragm is folded up concertina-fashion.

The field direction produces sideways motion instead of backwards-and-forwards movement, thus causing the diaphragm folds to expand and contract. Air is thereby sucked in and squeezed out, the particle velocity being much greater than the motion of the diaphragm that caused it, like a pip squeezed between the fingers.

Mechanical impedance of the driver is thus more closely matched to that of the air, hence the designation 'transformer'. So far the idea has been used for tweeters and mid-range units, but not for bass drivers.

Ribbon tweeter These work in the same way as a ribbon microphone in reverse. The ribbon is a ribbed or corrugated piece of aluminium foil suspended edgeways between the poles of a powerful magnet. Signal current passed through the ribbon sets up a magnetic field that interacts with that of the magnet to produce relative motion. Impedance of the ribbon is a fraction of an ohm, so a transformer is essential to match it to the amplifier. The ribbon is wider than those used for microphones, but there is a limit to its size; if too wide, the pole pieces are farther apart and the magnetic field is weakened. To increase its efficiency and provide an acoustic loading to limit its excursions, the ribbon works into a horn. The lowest frequency it will reproduce is governed by the size of the horn, but coloration can occur at that point so it is usually operated above it in conjunction with a crossover and bass unit.

Piezo tweeter Various substances have the property of changing dimensions if subjected to an applied voltage. Rochelle salt is a natural crystal which was once used, but is fragile and adversely affected by humidity changes. Manufactured ceramic materials such as lead zirconate titanate have now replaced it, being more robust and stable.

If two slices of the material are cemented together they form a *bimorph* that bends with the applied voltage, thus giving more movement than the single *monomorph*. Movement is still small and a cone or horn is appended to give a more efficient air coupling.

A feature of the piezo tweeter is its impedance which is very high at low frequencies, dropping to around 1000 Ω at 1 kHz, from where it falls smoothly to some 30 Ω at 40 kHz. The practical effect of this is that it can be used directly across the bass speaker without a crossover network, which is a major advantage.

One effect of the crossover network is that it isolates the tweeter from the amplifier so reducing the damping of spurious cone movement. The piezo tweeter, minus a crossover, is fully damped and so has a smooth response which is typically ±2 dB, 3–30 kHz.

Having no coil, the cone has much less mass than a moving-coil tweeter and so has a superior transient response and an output up to 40 kHz. Distortion is 0.5% over most of its range rising to just over 1% at 1 kHz. The load is mainly capacitive. Continuous signals of 15 volts (28 watts into 8 Ω), or intermittent music peaks of 30 volts (110 watts into 8 Ω), can be handled.

Plasma tweeter An electric arc generates high-temperature localised heat which produces violent air expansion. If the voltage producing the arc is modulated, so will be the heat and the rate of air expansion. Sound can thereby be produced, an example being the thunderclap. Several tweeters have been produced using this principle of which the plasma tweeter is a recent example.

The unit consists of a sphere of fine wire mesh at the centre of which is a fine needle electrode. A crystal-controlled oscillator produces a radio frequency of 27 MHz which is amplitude modulated by the audio signal and applied to the needle. The sphere is at earth potential, so a corona discharge takes place between its inside surface and the needle. Positive and negative ions are produced alternately and a high temperature of 1500°C appears around the needle causing violent expansion of the air within the sphere.

Modulating frequencies vary the ionisation, hence temperature rises almost instantaneously so the air expands and contracts in sympathy around the 1500°C mean. The result is a spherical sound wave which radiates in phase in all directions. As it is air itself that initiates the wave, there is no low-efficiency air coupling and no moving parts to add mechanical resonances and spurious motions. The in-phase, omnidirectional radiation eliminates the beaming and interference problems experienced with all other h.f. reproducers. The frequency coverage is 4.5–150 kHz. Total harmonic distortion is less than 1% at any point in its frequency range and maximum output is 95 dB at 1 metre.

Although the temperature is very high at the needle, the amount of heat is not large so the sphere does not get hot, and there is no danger. The unit is completed by a filter so that it can be connected across an existing speaker to extend its range.

Crossover networks

Use of multiple moving-coil drivers to handle different frequency bands necessitates filter circuits to separate them. These consist of inductors and capacitors arranged in low-pass, high-pass, and band-pass circuits, classified as: first *order*, having a 6 dB/octave attenuation; *second order*, having 12 dB/octave; *third order* with 18 dB/octave; and more rarely *fourth order*, with 24 dB/octave. An easy way to recognise the basic circuits is that first order has one component, second order has two, third order has three, and fourth order has four. Another description given to the second, third, and fourth order filters is

L network, T network, and π network respectively after the circuit resemblance to those letters.

Crossover frequency   The point where the bands overlap is the crossover frequency at which the response of each is at –3 dB. Because both drivers are producing output the total response is flat. In the case of the first order tweeter filter the current leads the voltage by 45° and with the bass filter it lags by 45°. The difference cancels acoustically to produce an in-phase output.

Higher order characteristics   The phase differences are greater with the higher order filters and do not cancel. With the even orders, they are up to 180°, necessitating the reversal of the tweeter connections to avoid signal cancellation at the crossover frequency. This also changes the phase relationship between low and high frequencies. Higher order filters are more prone to 'ringing': the storing and release of energy on transients, as can be revealed by tone-burst testing. They also need close tolerance components to avoid response anomalies in the overlap region.

Reasons for using high orders   Higher orders are used in spite of their disadvantages because first order filters have too gentle a slope and too great an overlap. If the filter roll-off coincides with the natural roll-off of the driver it is augmented and made steeper. One driver thereby has a steep slope and the other a gentle one, so producing a dip at the crossover frequency. Having a steep filter slope removed from the natural roll-off prevents this. With a first order filter the tweeter can be driven at too high an amplitude too far into the bass region, resulting in distortion and possible damage. Driver resonances can more easily be avoided with a steep slope.

The input of each network is connected in parallel across the amplifier output so that the same voltage is applied to all.

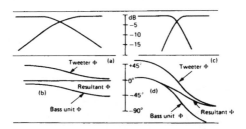

Figure 106. (a) first order crossover; (b) first order phase angle; (c) third order crossover; (d) third order phase angle.

252

Basic crossover circuits

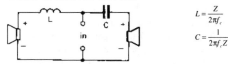

$$L = \frac{Z}{2\pi f_c}$$

$$C = \frac{1}{2\pi f_c Z}$$

(a) first order 6 dB/octave crossover.

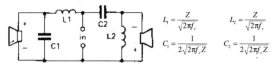

$$L_1 = \frac{Z}{\sqrt{2}\pi f_c} \qquad L_2 = \frac{Z}{\sqrt{2}\pi f_c}$$

$$C_1 = \frac{1}{2\sqrt{2}\pi f_c Z} \qquad C_2 = \frac{1}{2\sqrt{2}\pi f_c Z}$$

(b) second order 12 dB/octave crossover, L filters.

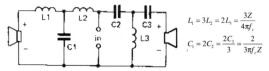

$$L_1 = 3L_2 = 2L_3 = \frac{3Z}{4\pi f_c}$$

$$C_1 = 2C_2 = \frac{2C_3}{3} = \frac{2}{3\pi f_c Z}$$

(c) third order 18 dB/octave crossover, T filters.

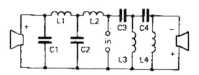

(d) fourth order 24 dB/octave crossover, π filters.

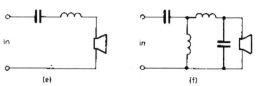

(e) First order band-pass (mid-range) filter; (f) second order band-pass filter.

Figure 107. Crossover circuits (f_c is the crossover frequency). These basic circuits assume that the driver impedance is constant, which is it not: it varies with frequency and resonances. They should therefore be considered as a starting point for a design that takes these factors into account; and can be modified to compensate for frequency irregularities in the drivers. Off-the-shelf crossover units are unlikely to be satisfactory as each should be tailored for a specific combination of drivers and enclosure.

Single full-range drivers

The crossover network has many disadvantages. Here are a few:

Crossover ringing   If a signal applied across a capacitor ends abruptly, a charge is left which then discharges through any parallel circuit. In the case of a crossover filter this will be the associated driver. A similar effect occurs with an inductor, an abrupt cessation of current causes a rapidly collapsing magnetic field which cuts its own windings. This generates a voltage that produces a current in any parallel circuit which will be of the same polarity as the original. So the effect is to sustain the current after the original has ceased.

The result is to blur any terminating transients in the applied signal, but this is not all. The induced current recharges the associated capacitor and when this ceases, it discharges back through the inductor. When this current in turn ceases, the collapsing field generates a further voltage which again recharges the capacitor. Thus the process is repeated to produce an oscillatory wave train which dies out only when the circuit losses reduce it to zero. The effect — which is called ringing — can clearly be seen on an oscilloscope when a tone burst is applied. It is especially pronounced around the crossover frequencies.

Effect on negative feedback   Ringing due to the crossover filter, being present at the amplifier output terminals, is fed back with the signal along the negative feedback path to a previous stage. However, as the spurious waveforms did not originate in the amplifier, there is no cancellation within the feedback loop, so they re-appear amplified at the output. They then pass to the drivers through the reactive crossover filter which delays them. So the spurious signals appear at the drivers, amplified and in random phase.

The precise effect depends on the nature of the feedback loop and the crossover filter. This probably accounts for the fact that some loudspeakers sound better with some amplifiers than others.

Cone damping   The springy cone suspension allows the cone to oscillate when an applied signal ceases. This effect is greatly reduced by the amplifier output damping. Damping factor is often quoted as a measure of how well spurious cone oscillations are thus suppressed. With an output impedance of

8 Ω and a resistance of 0.1 Ω, the damping factor would be 80, which is a typical value.

The amplifier damping also helps to tame cone and cabinet resonances, a factor not often considered. At resonance, the cone excursions are greater than at other frequencies, hence also its generated back e.m.f. This produces greater back current which in turn generates a greater opposing magnetic field around the speech coil. So damping produces a greater reaction at resonance and thereby tends to flatten response peaks.

However, the series elements of the crossover filter interpose an impedance between the driver and the amplifier. This makes nonsense of the 0.1 Ω typical amplifier damping resistance, as the impedance so introduced is many times that. Furthermore as the impedance is frequency selective, so also is the effective damping.

Phase distortion   As already shown, the phase difference with even order networks is 180° at the crossover point, necessitating reverse connection of the tweeter. The difference deviates from 180° on the overlap regions on either side of this point so giving rise to phase distortion.

It is thus evident that the signal suffers many forms of degradation as a result from the use of crossover filters. Added to these is another ill effect that arises from the use of more than one driver.

Spatial phase distortion   The sound waves generated by two drivers in the overlap region, mutually interfere. There are alternate regions of cancellation and reinforcement along a line at right angles to the line joining the centres of the two drivers. As they are usually mounted vertically, the main line of interference patterns will be horizontal although it is diffuse. The precise effect depends on the relative positions of the drivers and the listener, and the pattern varies with frequency. The general effect at any one listening point is of uneven frequency response and poor stereo resolution due to phase displacement.

At one time it was customary for makers to pack loudspeaker cabinets with as many drivers as possible in order to satisfy public demand, thereby compounding all the above effects. This practice has now virtually died out, and most modern loudspeakers contain only two drive units.

Eliminating the crossover filter   The crossover filter can be eliminated by using a piezo tweeter. Being capacitive, it has a rising impedance at low frequencies, and so can be shunted across the bass unit without fear of it being damaged by large amplitude bass. Having a lower mass, it responds to transients better than a moving-coil unit, and also has lower distortion.

Another solution is to divide the frequencies electronically at the pre-amp, amplify the two bands separately to then feed the two drivers. This eliminates all the problems of the crossover filter, but is expensive as it doubles the number of amplifiers required.

However the problem of spatial phase distortion and interference remains with both these options.

The full-range driver   The ultimate solution is to use a single unit to cover the whole frequency range. This may seem a backward step to the days of lo-fi record players, but if full range units are used in well-designed enclosures, excellent results can be obtained.

Controlled flexure, as previously described, enables the central portions of a cone to move independently of the rest and thereby serve as a very effective tweeter. A bonus with this arrangement is that the rest of the cone serves as an efficient 'surround' to absorb waves travelling outward from the cone centre. With a conventional small-coned driver, part of these are reflected back to produce standing waves. Upper frequency response can extend to 20 kHz with pin-point stereo definition as there is no phase distortion, as it does with the Kapellmeister* speakers, designed by the author.

---

* Fully described with constructional details in *An Introduction to Loudspeakers and Enclosure Design* published by Bernard Babani.

Surround sound and stereo

Direction location

In order to understand how stereo and the various surround sound systems work we need to know something of how we locate the direction of sound sources from any direction using only two receptors (our ears).

Time delays   There are several factors that are involved. The first is time delays and subsequent phase differences in the sound received by the two ears. A sound arriving from a source at one side arrives at the nearest ear about one millisecond ahead of its arrival at the opposite ear. The delay becomes less as the source swings to the front and is zero at dead front. For mid and high frequencies there are large phase differences, and the brain identifies the leading one as the source side.

With pure long-wavelength low frequencies, the phase difference is small so it is hard to identify the source direction. However, there are few naturally occurring pure low-frequency sounds, most include higher harmonics which serve as the direction indicators.

Amplitude difference   There is virtually no difference in amplitude between sounds reaching the two ears from a true single source unless it is very close to one ear. However, with stereo reproduction there are two sources; if both are reproducing a single sound, and one is louder than the other it appears that the sound source is nearer that side. This is an entirely artificial effect. It was the principle of the balance control in early stereo amplifiers, and 'pan-potting' as used during recording.

Head masking   Low-frequency sounds are diffracted around an obstacle which is small compared to their wavelength, but it casts an acoustic shadow when it is large compared to the wavelength (see pages 22–3). Thus the sound reaching the far ear has less high-frequency content than that arriving at the near ear, due to the masking effect of the head. This effect also is used by the brain as a direction indicator

These three factors serve to identify sound source location in a horizontal plane but do not distinguish between sounds coming from the front or rear, nor do they convey information as to the height of the source, whether above, below or horizontally level.

This is achieved by the outer ear or auricle. Some of the received sound passes directly through the ear orifice down the ear canal to the ear drum but some is reflected from the convolutions of the auricle. This produces complex interference patterns that differ according to the angle of the arriving sound. Reflections and difractions from the side of the head especially from the front, are directed into the auricle.

An object is an efficient sound reflector only when its size is greater than a quarter wavelength of the incoming sound wave. So the auricle reflects sound of frequencies above around 1.4 kHz, and the head extends this downward to about 450 Hz. Below this there is no spatial coding and height and front/rear perception is poor. However, as with head masking, most low-pitched natural sounds can be located accurately because of the presence of harmonics.

It follows from this that vertical spatial information can be decoded by the brain from only one ear although it is restricted to one hemisphere.

Stereo Stereo recordings are made by using two coincident directional microphones. This roughly approximates the position of the two ear drums; phase differences result which are similar to those present at the ears of a listener at the event.

An alternative is to use two microphones placed apart in the approximate position of the reproducing loudspeakers. The speakers then produce the sound fields as 'heard' by the two microphones. The coincident pair is now the most favoured method.

Other 'spot' microphones are frequently used to highlight certain instruments or soloists. The output from these is 'pan-potted' that is eventually mixed into the two stereo channels with amplitudes in each proportional to the position of the spot microphone. If proportional delay could also be introduced, it would give phase differences and a more natural sense of location than with amplitude difference alone.

To achieve an accurate stereo image in reproduction it is all important that there is no phase distortion. As most multi-driver loudspeakers produce phase distortion, poor stereo is all too common. Few stereo listeners experience the pin-point locational accuracy of which their recordings are capable of providing, because of this.

Even with phase-accurate loudspeakers only a flat horizontal spatial image is possible with a conventional stereo system, but there have been many attempts to enlarge it.

Ambience   In addition to the signal produced by the instruments in an orchestral recording, reverberation caused by sound reflections from the walls and ceiling of the studio or concert hall is also picked up by the microphones. When heard by a listener at the event, the effect is one of spaciousness and enhancement of the musical sound, but this is not achieved when heard through a two-channel stereo system, because it is all reproduced from the narrow frontal stereo stage.

Haffler effect   A simple way of extracting reverberation from the stereo signal and using it to produce a rear ambience depends on the fact that it is of random phase compared to that of the two direct stereo signals, and that much of the two direct signals are either in phase or nearly so with each other.

If two rear-mounted loudspeakers are connected in series across the two positive terminals of the stereo amplifier, all in-phase signals are cancelled, and only those that are out of phase are reproduced. Thus reverberation components of the composite stereo signal are isolated and reproduced from the rear loudspeakers. The arrangement is not perfect but can give good results. Much depends on the amount of reverberation present in the signal; it is usually greater for live performances in concert halls than for studio recordings.

Sansui QS 1   This was a reverberation extractor brought out around 1970, and was one of many produced at that time in an attempt to improve on the straight Haffler connection. The subtraction of the two stereo signals was effected as with Haffler, but the signal to one rear speaker was shifted by 90° at certain frequencies while that to the other was shifted by 90° in the opposite direction. Various effects and alteration of the frequency response of the rear channels could be made by a selector switch to give the best results from any particular recording.

Matrix four-channel systems

In the early 1970s quadraphony was introduced. There were two matrix systems, the SQ produced by CBS, and the QS developed by Sansui. The problem was how to contain four channels in a record groove so that it could also be played on an two-channel player.

The front and rear signals were added, subtracted, and phase shifted in a matrix. In the decoder, the process was reversed to produce cancellation and reinforcement. Thus the four channels were recovered.

The snag was poor channel separation. It could be made high between two channels but at the expense of the others, between which it could be as low as 3 dB. While poor separation with subsequent poor location definition may be unimportant for musical reverberation, there were many gimmicky recordings, in which the listener was in the centre of things and instruments were placed all around. For these good separation was important. Two-channel stereo compatibility was also poor.

The two system matrices differed in their mathematical relationships so that different channels were the low separation ones in the two systems. If an SQ record was played through a QS decoder or vice-versa, separation was poor for all channels, so they were not mutually compatible.

CD-4 system This followed very shortly upon the two matrix ones. All four channels were discrete and so had maximum separation all round. Produced by JVC, it was supported by RCA, and Philips (Japan).

A front plus rear channel for each side was recorded up to 15 kHz as a normal audio signal. In addition, a frequency-modulated carrier centred on 30 kHz was recorded which carried the front channel minus the rear. Deviation was −10 kHz +15 kHz, giving carrier frequencies between 20 kHz and 45 kHz. The two sides thus each consisted of the audio F+R and an f.m. F-R carrier, these being recorded on the two groove walls at 45° as usual. The system was thus similar to that of stereo broadcasting but without the pilot tone.

The amplitude of the f.m. carrier was controlled by the audio signal so that it was always 19 dB below the audio level. This prevented the carrier running at full amplitude when there was little or no signal. Had this not been done, variable groove spacing normally used to economise in disc space, could not have been used. A cartridge with a response up to 45 kHz and a Shibata stylus (see page 71) was needed to track the record.

The public adopted a wait-and-see attitude to see which system would become the standard; no-one wanted to invest heavily in equipment which would soon become obsolete. None of the systems would give way to the others, so all three perished through lack of sales, and quad died.

One possible factor in their demise was that they were not truly surround sound or *pantophonic* as they were claimed to be. When the four loudspeakers were placed in a rectangular configuration it was supposed to result in sound not merely coming from four directions but all points in between, just as it seems to come from all points between a stereo pair. The angle made by a pair of stereo loudspeakers and the listener should not exceed 60° otherwise there is a hole-in-the-middle effect. If the front quad loudspeakers are so spaced, and the rear ones likewise, there is bound to be a much larger angle at the sides, consisting of 360–120=240° for both, or 120° each side. The result was a hole-in-the-sides effect.

NRDC Ambisonic system   The brainchild of Professor Peter Fellgett of Reading University, it was announced in 1974. It was claimed to be *periphonic* as distinct from pantophonic, that is it reproduced sound with a vertical component as well as a flat 360° horizontal field.

With conventional stereo, sounds from the front and the rear are present but cannot be distinguished. If a third pressure microphone is added to two velocity left and right-hand microphones, it can be used to compare the phase of incoming sounds with those of the velocity pair and thereby establish whether the sound is from the front or rear. Thus pantophonic sound is possible with only three channels. For periphonic sound another velocity microphone is required to receive the vertical component.

Four Calrec microphones were used with the NRDC system, but not to feed four normal audio channels. As sounds are present in all four microphones but differ in amplitude and phase, each served as a reference to which phase and amplitude of signals in the other three was compared. The four were configured at the four corners of a tretrahedron, that is three-sided pyramid which has no corner diametrically opposite to any other.

The four signals were mixed in a kernel matrix, which processes an infinite continuum of variables, and the respective directions were defined by two parameters: horizontal, by degrees clockwise from North; and vertical, by degrees elevation above the horizontal. These were then coded as amplitude and phase differences between two signals recorded on a two-channel recording system.

For replay, the phase and amplitude differences were decoded back into the two direction parameters which then controlled the kernel matrix that apportioned the signal at appropriate levels to the various loudspeakers. Any number could be used but four was the minimum. Ideally these would be in the same positions as the microphones, but as a tretrehedral configuration was impractical in a living room, the compromise was to mount two loudspeakers in opposite corners at floor level, and the other two in the remaining opposite corners at ceiling level.

The system was ingenious but perhaps too complex for the time; the loudspeaker placing was not one which would recommend it to many potential users, and although pushed by the NRDC, it was never taken up by manufacturers.

Matrix H   As interest in domestic four-channel audio was dying, around 1977 news came that the BBC were experimenting with their own quad matrix system to broadcast with their stereo transmissions. It was called Matrix H, but no details were released and it too slipped into limbo.

Devices to produce rear ambience continued to appear for a while. These either abstracted reverberation from the stereo signal or generated artificial reverberation. There was a 4-channel resolver by Neal, and a grandly named *audio pulse digital time-delay reverberation system* by Pyser. However, it seemed that the public in general felt that finding room for an extra two speakers wasn't worth the limited appeal of rear ambience. So interest in domestic surround sound faded.

Dolby stereo   Surround sound persisted in the cinema with the Dolby system. Four channels fed four loudspeakers in a diamond configuration, one centre front, two at the sides and one at the rear. The two side channels were straight stereo, and the centre one a mono combination of the two to fill the hole-in-the-middle for those sitting at the front centre. The rear channel was delayed by between 30–100 µs to ensure that those sitting at the back heard the signals first from the front.

Some t.v. receivers now have a decoder and detachable loudspeakers so that the surround effect can be obtained with Dolby stereo films. As there is no danger of a hole-in-the-middle effect with t.v. sets, the fourth loudspeaker is assigned to accompany the third at the rear.

Zuccarelli holophony   In 1983 the real breakthrough occurred and a series of demonstrations in London by inventor Hugo Zuccarelli astonished cynical sceptics and hard-bitten audio journalists. From two-channel headphones they heard sounds seemingly coming from all around, overhead and beneath. In a darkened room some found the effect terrifying. Remarkably, when one earphone was removed, the directional effects continued although reduced.

The secret was the use of a dummy head for recording. Nothing new in that, but hitherto dummy heads with microphones for ears, have merely been used to add head-masking. Zuccarelli's dummy head — named Ringo — was far more than that. The auricles were accurately shaped like human ears. Ear canals were formed with elliptical tubes 24 mm long, with a quarter-turn twist and an abrupt dilation halfway along; the first 8 mm was soft plastic and the rest hard plaster to simulate fibro-cartilaginous and bony sections respectively.

At the end of each canal was a microphone with a 7 mm diaphragm, orientated in the same plane as the ear drum. From the back of each microphone a tube ran down to the oral cavity to simulate the Eustachian tubes which run from the back of the ear drum to the pharynx. As hair was found to have an effect on masking, Ringo was also provided with a wig. Thus all the natural spatial encodings were provided.

Double encoding resulting from both Ringo's auditory system and the listener's, may have been expected but did not occur, because the listener's source was a stationary earphone. So the spatial information was only that of Ringo's. Likewise double resonances were avoided. The human auditory system has resonance peaks at 400 Hz, 4 kHz and 12 kHz, Ringo's was at 2.5 kHz and 7.5 kHz, but these were notched out during recording.

As all the spatial encoding was acoustical, the microphones fed a normal stereo pair of channels. No decoder was needed for replay only a conventional stereo player and headphones, the system was thus perfectly compatible with stereo and mono systems, some spatial effects could even be experienced with the latter, as was demonstrated.

There were two possible snags. Headphones are not popular for general listening. Loudspeakers could be used providing they had no phase distortion, which most have. Single-driver units such as the Kappelmeister would have been suitable. However, leakage of sound from each loudspeaker to the opposite ear, and wall reflections could degrade the effect. The other snag was the inability to use pan-potted spot

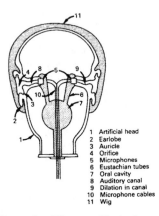

1	Artificial head
2	Earlobe
3	Auricle
4	Orifice
5	Microphones
6	Eustachian tubes
7	Oral cavity
8	Auditory canal
9	Dilation in canal
10	Microphone cables
11	Wig

Figure 108. Dummy head Ringo used in the Zuccarelli holophony system.

microphones, but a re-think on studio acoustics and instrumental positioning could have eliminated the need for them. Apart from some records by the group Pink Floyd, the system unfortunately failed to interest the recording industry.

Sensaura   A 'new' surround sound system from EMI's Central Research Laboratories was announced in October 1993. CRL are keeping the details quiet at the time of writing, but it appears to be a copy of Zuccarelli's holophonic system plus digital signal processing. A dummy head is used plus torso but minus wig. The auricles are replicas of human ears, and nose and mouth apertures suggest internal cavities similar to Ringo's. The microphones and head construction was supplied by Bruel & Kjaer. As with Zuccarelli's system, the head's auditory canal resonances are notched out.

Pan-potting is provided for by digital mixing of the additional signals in proportion to the microphone's position relative to the head. Whether the extra signals are merely amplitude potted, or proportionally delayed to each channel as well to produce the vital phase differences, has not been revealed. Certainly, they do not contain the spatial coding provided by the head auricles and auditory canals, which are the system's main features. So pan-potting is likely to degrade the spatial effect, even though the system has the capability of conveniently handling it.

The signals are converted to digital by a 20-bit analogue-to-digital converter, and processing and mixing is carried out at 24-bit resolution.

Roland Sound Space   This is another recent system, for which information is sparse. It is said to simulate a dummy head electronically, in the final mixing of the sound tracks, and so appears to be a superior type of pan-potting. If it simulates all the spatial encodings of the auricles and auditory canals and not just the obvious ones of delay and masking, the degrading of spatial information by ordinary pan-potting would be avoided, and it would appear to be the ultimate system. However in the absence of positive information, it would seem unlikely that it does this.

Stereo image extension   It is sometimes noticed that inter-station f.m. white noise appears to extend beyond the stereo loudspeakers. The ears are fooled by the random phase differences. Using the same principle an image widening can be achieved by reversing the phase of one channel. However this also results in poor bass due to partial low-frequency cancellation. A practical solution is to reverse or change the phase of only the high frequencies in one channel. This is now done in some portable stereo recorders.

Public address

Objectives

Intelligibility The first and most important requirement of a
p.a. system is intelligibility. Speech must be heard with clarity,
the syllables distinct, and words readily understood.

Adequate volume Listeners must be able to hear
comfortably without strain or being deafened. In an auditorium,
peaks of around 75 dBA are required, for outdoor events
80 dBA. Where there is high ambient noise such as in a
machine shop, the level must be 10 dB greater than the noise,
but this could take it above the safe hearing limit (see page 4).
High level by itself accomplishes nothing if speech is
indistinct, so volume comes second to intelligibility.

Acoustic stability Closely related to adequate volume is
acoustic stability. An unstable system limits the volume
available and is also liable to run into feedback with little
warning. Various factors affect the stability and so must be
considered when planning a system.

Noise Electronic noise is generated by all amplifiers and
mixers, but it should be of such a low order that it is not
audible to any part of the audience. This is not difficult to
achieve with modern equipment. Mechanical noise arises from
microphone handling and static picked up on cables.
Professional microphones have internal shock absorbers to
reduce handling noise, although the directional models
preferred for p.a. work are more susceptible to it. Muting
switches on the control desk are useful to silence the channel
when microphone stands are being adjusted. Cables having a
slightly conductive covering eliminate static noise.

Naturalness A highly desirable though not essential quality.
In auditoria it can be achieved by avoiding resonance peaks,
especially in the microphones; not tailoring the frequency
response too drastically; using a limited number of column
speakers positioned so that the sound comes from the direction
of, and on a level with, the platform; generally using high-
grade equipment. The perfect p.a. system is one whereby the
whole audience can hear without being aware that a system is
in use.

In noisy situations it may have to be sacrificed to achieve
intelligibility by using noise-cancelling microphones and
frequency tailoring.

Reliability   A major breakdown in a large auditorium with a capacity audience is a disaster. The system should be designed to be inherently reliable, with standby equipment. and monitoring devices at various points to locate faults rapidly should they occur.

Visual appeal   A temporary set-up that has wires looped about and left loose over the floor looks unprofessional and could constitute a safety hazard. It takes very little extra time to clip them neatly into place. Tarnished stands, dented microphones and badly scratched speaker cabinets also detract from the appearance, and the installer's reputation.

Loudspeaker distribution

Low impedance losses   If long cables are used to supply a low impedance loudspeaker their resistance can form a significant part of the total load impedance. A resistance equal to the loudspeaker impedance means that about half the power is dissipated in the cable, while paralleling several speakers from it could lower the impedance to less than the cable resistance, resulting in an even greater power loss. Parallel connected speakers could also lower the load impedance below the minimum rating of the amplifier. Series connection avoids this, but is inconvenient to wire, and switching off one speaker affects all the others.

High impedance   The solution is to use high impedance amplifier output and speakers. Cable resistance thereby becomes negligible compared to the speaker impedance. Any number of speakers can be paralleled providing their combined impedance is not less than that of the amplifier.

A transformer in the amplifier steps up the impedance and individual transformers in the speakers step it down to the impedance of the driver. They are tapped so that different power ratings can be selected, and thereby different volume levels, to suit the environment.

100 volt lines   Calculation of the respective impedances is an awkward way of determining the load, so the output is designated in voltage. 100 V being the standard, though 70 V is also used. This means that with a sine wave at full-rated power the amplifier delivers 100 V to the load. Normally the signal voltage level varies and is much lower than that. Calculating the amplifier loading thus becomes a simple matter of adding the wattage tappings of the speaker transformers.

Transformer tappings   A speaker with a particular wattage tapping will take the same power and deliver the same sound output from the 100 V line output of any amplifier irrespective of its power rating. The only difference is that a large power amplifier will supply more speakers than a small one. Higher volume can be obtained by increasing the power tapping on the speaker transformer. Note that a 100 V speaker tapping will take about half the power when operated from a 70 V line. Care must be taken that all connections to the transformers are phased so that all speakers working in the same area are in phase.

Table of transformer ratios

	Power at 100 V						
Impedance	1.25	2.5	5	7.5	10	15	20
3 Ω	51.6	36.5	25.8	21	18.3	14.9	12.9
4 Ω	44.7	31.6	22.4	18.3	15.8	12.9	11.2
8 Ω	31.6	22.4	15.8	12.9	11.2	9.11	7.9
16 Ω	22.4	15.8	11.2	9.1	7.9	6.5	5.6
20 Ω	20	14.1	10	8.2	7.1	5.8	5.0
24 Ω	18.3	12.9	9.1	7.5	6.5	5.3	4.6
	0.625	1.25	2.5	3.25	5	7.5	10

Power at 70 V

Table of amplifier output impedances at various powers

Power	100 V	70 V	Power	100 V	70 V
20 watts	500 Ω	245 Ω	50 watts	200Ω	98 Ω
25 watts	400 Ω	196Ω	75 watts	133 Ω	65 Ω
30 watts	333 Ω	163 Ω	100 watts	100Ω	49 Ω
40 watts	250 Ω	122Ω	200 watts	50Ω	24.5Ω

Avoiding feedback

When a microphone and loudspeaker are working in the same volume of air, there is a coupling between them; sound from the speaker reaches the microphone even though it may have come via many reflections and be greatly attenuated. This is re-amplified, emerges from the speaker louder, is picked up again by the microphone and amplified further. The cycles continue rapidly until they build up to a howl.

There is a critical point at which the system becomes unstable. Below it, although feedback occurs, the amplifier gain is insufficient to overcome the coupling losses for the howl to build up. Just below the critical point, *ringing* occurs, and decaying feedback is triggered by every sound picked up by the microphone. A poorly designed system often operates close to this region to obtain sufficient volume, but intelligibility suffers. In such case it is better to reduce the volume so that what there is can be understood. Ideally, the system should be operated well below critical feedback point and still have ample volume, but this is not always easy to achieve in practice.

Microphones Microphone choice has a major effect on feedback. Two factors are significant; the *polar response* and the *frequency response*. Unlike an omnidirectional microphone, a cardioid or hypercardioid can be angled toward the source and away from the direction of the speakers. But even at a random angle, the feedback is less because the cardioid picks up 4.8 dB less reflected ambient sound from the speakers, and the hypercardioid 6 dB less, than an omni. Extra gain of a similar order is therefore available before critical feedback point is reached.

Frequency response peaks indicate higher gain at those points. They will thus initiate feedback even though the rest of the response is well below the critical feedback level. Additionally, feedback starts readily at a peak. If the response is flat, the whole curve can operate closer to the critical point without exceeding it and feedback is sluggish in building up. So a hypercardioid with a very flat response is required. Among the very few models apart from capacitor microphones that fulfil these conditions is the Beyer M260 N80 ribbon.

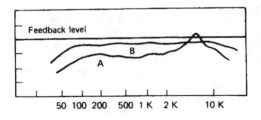

Figure 109. Microphone A's peak exceeds feedback level but microphone B's flat response enables it to be used at higher gain without feedback.

Acoustic absorption Stage or platform microphones are normally angled toward the participants, which means they are facing the wall behind the stage. This reflects sound from the auditorium into the fronts of the microphones and thereby encourages feedback. A considerable improvement can be obtained by acoustically deadening it. The easiest and most visually acceptable way of doing this is to hang a heavy curtain in deep folds over the whole length of the wall.

Loudspeakers Feedback can be greatly reduced by the use of column loudspeakers providing they are positioned and angled correctly. It is remarkable that in the majority of commercial installations, column loudspeakers are incorrectly installed. They are usually mounted high up either side of the stage with little or no forward tilt, thereby beaming the sound over the heads of the audience to hit the back wall and be reflected back to the stage. The result is early feedback and an indistinct, reverberant sound.

The correct position is much lower, with the bottom of the speaker at about audience shoulder height, and tilted slightly so that the top is in line with the head of someone standing at the back of the auditorium. An inward angle brings the central regions of the front rows into range and minimises reflections from the side walls.

The angle should not be so great that the front of the column is visible from the platform. For larger halls, a second and even third pair of speakers can be used at roughly a third and two-thirds the distance from the platform to the back of the hall. The inward angle of these should be progressively greater; as a rule of thumb, aim each speaker at the opposite back corner of the hall. The sound is thus beamed into the audience where it is required and little is reflected directly from the walls.

If a higher mounting is used, the angle of tilt must be greater, but the range of forward coverage is reduced by the angle. Where there is a rising floor toward the back of the auditorium, as with balconies, the columns should not be tilted because there is already an angle between the audience and the sound beam.

If in doubt as to the position in any situation, use the propagation pattern of Figure 105 to ensure that the sound is directed into the audience with the minimum overspill.

Equalisers   These are sometimes used to reduce peaks in the system and thereby increase the available gain before the critical feedback point. They are set up by using a frequency counter. Starting with a level position on the equaliser, the gain is advanced until feedback occurs. The frequency is observed on the counter and the nearest equaliser control reduced until the feedback stops.

Gain is increased once more until feedback recommences at another frequency and this dealt with in the same way. The process is continued until all the resonances have been tamed. If a frequency counter is not available, the controls can be set by ear, and trial and error.

The drawbacks are that the resonant frequencies are unlikely to coincide with those of the equaliser especially if its bands are spaced at one octave or more. Furthermore, many peaks arise from interference reflections between the microphone and nearby objects. These can be drastically changed if the microphone is moved in use, thereby upsetting all the adjustments. The practical advantages are therefore very limited.

Frequency shifter   All frequencies passing through the device are raised by about 5 Hz. This has no audible effect, but with each feedback cycle, the frequency is increased and quickly ascends to above the range of the system. A howl cannot therefore build up. Instead, the effect is of a throbbing squeak, but it permits some 6 dB of extra gain before audible feedback occurs.

This can be very useful, but as the device is complex, provision should be made to switch it out and operate directly, should a fault develop. It should never be used to 'paper over the cracks' of a badly designed system. Use suitable microphones, position loudspeakers correctly, improve platform acoustics, then use a shifter to give a useful extra margin.

Delay lines

Delay lines are used to delay an audio signal by a fixed amount, and can be either digital or analogue

Digital delay   The signal is first compressed to reduce the dynamic range to one that can be quantised by a restricted number of digital bits, and after pre-emphasising the treble it is sampled at around 40 kHz. It is then converted to binary digital words and passed along a memory store. At the other end the binary data is applied to a digital/analogue converter, passed through a low-pass filter to remove the quantisation steps, de-emphasised, and the dynamic range expanded to that of the original. The output from the memory can be taken from earlier points so giving a smaller delay as required. These are not often used in p.a. systems because of their cost.

Analogue delay   These are built around the *bucket-brigade delay line* which consists of a chain of from 512–3328 capacitors contained in a single i.c. Several i.c.s can be cascaded to form longer chains if required. The signal is sampled at three times the highest signal frequency and each sample charges the end capacitor of the chain. On receipt of a pulse from a clock generator, the capacitor passes on the charge to the next capacitor, a further pulse transfers it to the third one and so on down the chain. It emerges at the end with a delay dependent on the number of stages it has passed through and the frequency of the clock.

A problem is that the samples come in succession, yet capacitor 1 cannot receive another sample while it is transferring the previous charge to capacitor 2. This is overcome by actuating alternate pairs. So capacitor 1 transfers to capacitor 2 while capacitor 3 transfers to capacitor 4. On the next pulse, capacitor 2 transfers to capacitor 3 while capacitor 4 transfers to capacitor 5. This means that only half the capacitors are conveying information at any time and the effective length is half the actual length. Alternate positive and negative pulses are required to trigger the odd and even numbered capacitors.

Charge transfer   As it is difficult to discharge a capacitor completely, instead of the actual charge passing from the first to the second one, the second is fully charged initially and then is discharged into the first, which already contains a sample. When the first is full, the second still has some charge left which corresponds to the sample in the first one when the transfer started. So a charge equal to the sample is left in the second capacitor, and an effective transfer has been made.

Output filter The signal appearing at the end of the line is a series of alternate pulses and gaps which would require a very steep, complex and expensive low-pass filter to smooth out. Instead, an output is taken also from the penultimate stage and combined with final one, The pulses of the one occupy the gaps of the other, giving a much smoother result which needs only a modest low-pass filter.

The delay obtainable is given by:

$$d = \frac{n}{2f_c}$$

in which d is the delay in ms; n is the number of stages; f_c is the clock frequency in kHz.

Clock frequency This must be at least three times the highest audio frequency so the required frequency response sets the limit on the delay available. For public address 6 kHz is a reasonable upper limit, so a frequency of 20 kHz may be considered a suitable minimum. However, many devices have 'hairpin' distortion/clock frequency curves. Distortion can be as high as 10% at 20 kHz and fall to 0.4% at 40 kHz, rising again rapidly above 60 kHz. For such devices the clock frequency should be kept within 30–80 kHz to avoid high distortion, preferably within 40–65 kHz.

Delays can be varied by means of the clock frequency within these limits, but some devices have a series of taps enabling smaller delays to be selected. Otherwise the device can be chosen to give the approximate delay required, and the exact amount obtained by choice of frequency.

Use of delay lines

Intelligibility If the distance from one speaker to a listener is greater by 50 ft (15.4 m) than that from another giving similar output, the resulting 45 ms delay or more will blur speech syllables and give poor intelligibility. If the sound pressure level from the distant speaker is much lower, the effect is slight, but the long range of the line-source speaker coupled with strong reflections, can give rise to problems of this nature.

The effect could arise in a long hall where sound from the front speakers reach the rear area along with that from the speakers there. P.a. speakers on railway station platforms are another common example.

A delay line feeding the amplifier which supplies the rear speakers can be used to add a delay similar to the acoustic delay, so that sound from both sources arrives at the affected area at the same time.

Source location In some situations such as for stage drama or concerts, it is desirable to make the sound system as unobtrusive as possible. In such cases the sound must appear to come from the stage and not from loudspeakers elsewhere in the auditorium.

Haas effect When sounds arrive from two or more spaced sources, they all appear to come from the direction of those arriving first, provided the delay is no more than 30 ms and the pressure level difference is no more than 10 dB. If a delay is introduced that is longer than the acoustic delay, sound from the front speakers will arrive at other parts of the auditorium before that from the local ones and all the sound will seem to come from the front. The level from the local speakers must be kept to within the 10 dB limit.

Table of distances and delays at 60°F (15.5°C)

Distance	Delay	Distance	Delay
50 ft 15.4 m	45 ms	80 ft 24.6 m	72 ms
60 ft 18.5 m	54 ms	90 ft 27.7 m	81 ms
70 ft 21.5 m	63 ms	100 ft 30.8 m	90 ms

Although digital delay lines are superior, analogue bucket-brigade devices are much cheaper and quite adequate for p.a. work. They may thus be used where cost would otherwise rule out the use of delay techniques.

Ceiling speaker mounting

Ceiling-mounted speakers are sometimes used when unobtrusiveness and local low-power coverage is required. Hotel foyers, dining rooms, corridors and similar situations can benefit from this type of installation. It is not recommended for halls and auditoria.

Spacing Propagation pattern from a ceiling-mounted speaker is that of a half-doublet, so the sound pressure level (SPL) at an angle from the perpendicular axis is:

$$SPL_2 = \cos\theta SPL_1$$

where SPL_2 is the sound pressure off axis; θ is the angle; and SPL_1 is the sound pressure on the perpendicular.

To achieve an even coverage of direct sound with a line or matrix of speakers each should be mounted over the point where the half SPL from adjacent speakers reach the audience. From the above formula half SPL is at an angle of 60°. The spacing between speakers is therefore the base of a right-angled triangle having a 60° angle at the apex and the length of the perpendicular being the height of the ceiling above the audience. This is obtained from:

$$d = \tan\theta h$$

in which d is the spacing; θ is 60°; and h is the height above audience. From this it is seen that the spacing should be 1.73 times the height of the ceiling above the audience.

Temperature Sound propagated from the ceiling must pass through various layers of air at different temperatures. The resulting refraction bends the sound waves toward the colder lower regions, with the effect of focusing the sound on to a smaller area beneath. Thus sound coverage becomes uneven as temperature builds up near the ceiling. The effect does not occur with sound propagated sideways, i.e. from column speakers.

Feedback The side propagation from a ceiling-mounted speaker makes the system vulnerable to feedback if the microphone is within range. A sound pressure of 0.1 of a speaker's maximum output is sufficient to cause feedback. From the above formulae this level is propagated at an angle of 84°, and the distance of this point from the speaker is 9.5 times the height of the ceiling above the microphone. So feedback could occur if a non-directional microphone is brought within this range.

Resonances While the fundamentals of the length and width modes are usually too low to come within the range of the system, the vertical resonant mode can be strongly excited by ceiling-mounted speakers. The result is a peak between 50–100 Hz, giving a boomy effect. It is not excited by the sideways propagation of column speakers. The vertical mode resonance is often reinforced by resonances in the ceiling cavity.

Multiple delay When a listener receives sound from a number of loudspeakers the effect is similar to hearing sound in a reverberant environment, and speech is rendered indistinct. The delays between the different units also result in reinforcement and cancellation at various frequencies so further adversely affecting the quality of the sound.

Natural location The sound of a voice coming from overhead and, in some locations, from behind, robs it of its naturalness. Thus ceiling-mounted speakers have many disadvantages and are only suitable for spot announcements in limited areas.

LISCA (line-source ceiling array)

The LISCA system — although ceiling-mounted — avoids all the disadvantages of the conventional matrix of ceiling-mounted loudspeakers described earlier, and is even an improvement on vertically mounted line-source columns.

Normally, a column should never be used horizontally for two reasons. Firstly, the sharp cut-off at the ends restricts coverage to a narrow band in front of it having a width equal to the column length. Secondly, the wide side dispersion then becomes vertical so sound is directed upwards to the ceiling and upper walls where it produces unwanted reflections.

LISCA consists of one or two horizontal line-sources mounted in the ceiling, but its proportions and angle avoids the first drawback by making a virtue of it, and makes good use of the second.

The line-source extends the whole width of the hall thus eliminating the problem from end cut-off, because the whole audience area is within its frontal region of propagation. The end cut-off actually minimises reflections from the side walls, so is an advantage.

The first line-source array is mounted a little forward from the first row of seats, with the loudspeakers tilted to an angle of 64° facing the audience. It behaves acoustically as a single source and the resulting plane pressure wave is directed downward and along the hall arriving at all points in the auditorium in phase and in unison. It thereby avoids the interference effects of multiple-source units.

Intelligibility is accordingly high, and is consistent, being the same at all points. The angle of propagation avoids strongly exciting the vertical floor-to-ceiling resonance. The overall result is speech of exceptional clarity everywhere.

Feedback The tilt of the loudspeakers of the first line-source array put them at an angle of around 90° to any microphones on the platform, so the direct sound field from them there, is cos 90° which is 0. Thus no direct sound can reach the microphones from the array. This is in contrast to the ordinary ceiling matrix which. by facing downward, radiates sound from its first row toward the platform.

As with all systems. there can be feedback from reflected sound from the auditorium, but even this is reduced as there is little or no sound directed at the side or rear walls, most being directed into the audience.

Sound levels and range The significant dimension affecting all calculations, is h, the height of the array above the seated audience-head-level which is about $3^1/_2$ ft or 1 m from the floor. Thus h is the floor-to-ceiling height minus $3^1/_2$ ft or 1 m. The first row of seats just beneath the first array, being 64° off-axis, receives a sound level of cos 64 = 0.44 times, which is –7 dB that of the same distance along the on-axis line.

The on-axis line converges with the audience-head-level at a distance from the array of: $h/\sin(90° - 64°) = h/0.44$ or $2.3h$. As the propagation loss from a line-source is 3 dB for a doubling of distance, the level along the on-axis line to the point where it reaches the audience is therefore about –3.5 dB.

So with –7 dB under the array at the first row, and –3.5 dB at the on-axis point, the latter is +3.5 dB compared to the former. Thus the first rows receive only enough amplified volume to reinforce the speaker's natural voice. As the speaker will probably be nearer the first row than the array, all the sound will appear to come from him due to Haas effect, thereby giving an entirely natural effect.

The floor distance from the first row to the +3.5 dB on-axis point is $h/\tan 90° - 64° = h/0.49$, or approximately $2h$. So $2h$ is the floor distance from the first row to the point where the sound pressure increases to + 3.5 dB, the maximum level.

Beyond this it declines as the line of propagation goes off-axis again and the distance increases. However, distance has less effect than may be expected. This is because the normal cylindrical expansion of the wave from a line-source in a large free-field environment, giving a 3 dB drop for a doubling of distance, is here restricted.

With LISCA, the wave expands semi-cylindrically, obeying the normal attenuation law until it fills the space between floor and ceiling. Beyond this no further expansion is possible as it is constrained by the floor, ceiling and walls. From there on the effect is like sound travelling along a tube, and in theory there can be little further loss beyond this point.

In practice there are losses due to absorption by the audience, carpet, if fitted, curtains and padded seating. The range limit beyond $2h$ will therefore depend on the furnishings. At $4h$ the propagation angle is narrowed to the point where the loudspeaker cones begin to be masked by the ceiling. Low and mid frequencies are diffracted but high frequencies are not, so intelligibility starts to deteriorate. Also, the off-axis angle increasingly reduces the level. Thus $4h$ can be considered the maximum range to give adequate sound level with high intelligibility. If the length of the hall from first row to last is greater than $4h$ a second line-source array will be necessary.

The second line-source

The second line-source should be located, not where the output from the first is tailing off at $4h$, but where it is strongest at around $2h$ or a little beyond at $2^1/_2h$. Here, the output from the second array which is immediately overhead, is -3.5 dB compared to that from the first. So, with most of the sound coming from the first, the perceived location is still forward and there is little awareness of the second array overhead.

Moving back from this point, the output from the first line-source starts to diminish while that from the second increases, thus maintaining an even sound level throughout. Furthermore, as the second array is now forward relative to the listener, the frontal natural source location is maintained.

In very long halls or those with a short h dimension, the distance from the proposed second line-source location to the last row of seats may exceed $4h$. In this case, the location of the second array may have to be taken back to $2^1/_2$–$3h$.

There may then be some awareness of an overhead source immediately beneath it because the output from the first is less than $+3.5$ dB at this point. However, the effect is local, and sound level and intelligibility are not affected. The situation where the length of a hall from the first row of seats to the last exceeds $6h$ would be unusual.

It may seem that having two arrays would create the mutual interference effects that characterise the ceiling matrix. However, at any location, either the volume levels are dissimilar, or the sound path difference between each array and the listener is large. For interference to affect mid and high speech frequencies, sound levels must be similar and path differences small. While these conditions exist in many locations with the ceiling matrix system they are found nowhere with LISCA, even though two arrays are used.

Pseudo stereo effect An unusual effect is often heard with LISCA systems. Not only does the sound come naturally from the front as it should, but it also seems to correspond to the location of speakers on the platform, either to left or right.

The explanation seems to be as follows: our ear/brain combination tries to identify the location of all sound we hear. It does this by comparing the arrival times of sounds received by the two ears. Those arriving at one ear before the other are perceived to come from that direction; also the brain judges the precise angle by the difference in time. This incidentally is why it is difficult to locate low frequency sounds, the wavelength is so long that there is virtually no phase difference from one ear to the other.

Nearby sounds are especially easy to locate because of their spherical wavefront, the curve ensuring that the sound to one ear is well in advance of the other. With more distant sounds, the curvature is much less, so location is less easy.

With LISCA, the wavefront is plane, there is no curvature at all, and it arrives at both ears at precisely the same time. So there is no lateral directional information. The brain then seeks to replace the missing directional information with data from another sensor; the eyes. So, when observing participants at any position on the platform, the brain signals its auditory section that that is where the source is.

It is an illusion, but then so is stereo. Nonetheless, it provides a useful bonus of extra naturalness, which together with the desired graded sound level from the front, the even coverage over the rest of the auditorium, the extremely high degree of intelligibility and clarity, and the low feedback characteristic, makes LISCA one of the most effective public address loudspeaker systems available.

(Note: A full description of the LISCA system with assembly and installation details is covered in *Public Address Loudspeaker Systems* published by Bernard Babani.

Mixers

The mixer is the central control of the system. All inputs are channelled into the mixer, which then feeds the power amplifiers. These need only to have a single high level input and are described as *slave* amplifiers. Many amplifiers combine a power section with mixing facilities, but these are usually limited.

Microphone inputs   The principal feature here is that the input should be of adequate sensitivity. High quality microphones, other than capacitors, have a low output and a sensitivity of 0.3 mV is generally required (a lower figure means a higher sensitivity). Many mixers are designed for microphones with higher output and so have low sensitivities. Low noise is important, but modern semiconductor technology has achieved acceptable levels even in inexpensive units.

Professional mixers use balanced microphone input circuits, but most non-professional units have unbalanced circuits. With these, the quasi-balanced circuit can be used to advantage. Capacitor microphones need a polarising power supply which must be conveyed via the input socket and cables.

Reverberation   One or more of the microphone channels may have artificial reverberation which improves the sound of mediocre vocalists by adding 'body' to the sound. There are three ways of providing this: a steel spring with a transducer at one end to feed the signal in and another at the other end to take off the multiple reflections generated; a bucket-brigade delay line with regenerative circuit to feed the output back to the input for multiple re-cycles: or a digital delay. A control is provided to vary the amount of reverberation to direct signal.

Gram inputs   An input with RIAA equalisation is required for the use of a magnetic pickup. With cheaper mixers, the gram is not equalised which means that a ceramic pickup is intended, but these often have too low an input impedance for such pickups. A 1 MΩ minimum impedance is required otherwise there will be severe loss of bass.

Crossfading   Where two gram inputs are provided, a cross fade control fades the one up as the other is faded down. It works as a straight volume control if only one input is used.

Tape/auxiliary   This is a straight input with usually 100 kΩ impedance and a sensitivity of around 0.5 V.

Tone controls   These differ in complexity from a simple top cut, through treble and bass cut and boost, to a built-in equaliser. There can be one overall control or separate controls for each input. For most applications an overall treble and bass cut-and-boost pair of controls is sufficient.

Monitoring   Output meters are generally provided, as sometimes is a headphone socket.

Prefade   This permits the audition by headphones of the gram channel with the fader down, so enabling it to be faded up on cue.

Output   The standard output is 1 mW into 600 Ω, which in terms of voltage is 0.775 V. This therefore is the 0 dB reference level. For a 200 Ω impedance the 1 mW output voltage is 0.447 V. Some mixers are stereo that can be switched to mono output.

Outdoor classical concerts

Outdoor classical concerts appear to be increasing in popularity. Given the sound levels generated by a symphony orchestra playing fortissimo, there may seem to be no need for electronic sound reinforcement. This is not so. There are many passages for solo instruments in classical works, and while instruments in the brass section may indeed be heard on their own, the more delicate sounds such as those of the oboe and cor anglais would be lost. Similarly, the string section as a whole may get through, but solo violin parts would not. One reason is the lack of the reflecting walls of the concert hall to concentrate the sound to the audience, and another is the larger audiences, often several thousand.

Providing an effective sound reinforcement for an outdoor classical concert requires skill and experience, but the basic principles here given should aid the engineer with limited experience to get started.

Discreet coverage   The idea of hearing a live orchestral performance through loudspeakers is anathema to most music lovers. To hear the real thing is their main reason to leave their hearth and hi-fi where they can listen for free and in comfort. So the installation, especially the loudspeakers, must be as unobtrusive as possible.

Loudspeaker location   It may be possible to conceal loudspeakers in trees or bushes. A few potted shrubs could be arranged for this purpose, and has the advantage of allowing the loudspeakers to be put in the best position, instead of using available cover which may not be in the ideal place.

Loudspeakers should not be placed at the sides of the audience but in a frontal position so that the sound is coming from the general direction of the platform, and they should not be so near the audience that they can be identified as the sound source.

Stereo is not normally feasible because those sitting at the side of a large audience would be too far from the opposite-channel loudspeaker to receive any sound from it. There could be an exception if a pair of loudspeakers were situated one on either side of the platform, and the front of the audience was some distance away. This distance would have to be greater than the spacing between the loudspeakers.

A good site location for the platform is in front of a small lake or pool with the audience on the opposite side. Air temperature inversion often occurs over water, which has the effect of directing sound waves downward toward an audience instead of their usual propagation pattern upwards (see page 24). This also keeps the audience from being too close and so may make a stereo set-up feasible. It also gives a very attractive appearance to the scene.

Delaying the sound   When two identical sounds arrive at a listener from different directions, the brain locates the source as that of the first to arrive, and in fact it appears to be the only source. This effect can be used to distract listeners from the loudspeakers. If the amplified signal is passed through a high-quality delay line to give a short delay, the orchestra will seem to be the only sound source. If the situation necessitates a loudspeaker location near to the audience, the delay must be sufficient for the live sound from the platform to arrive first (see table on page 272). A little added reverberation will improve the effect further by giving a more concert-hall like sound.

Type of loudspeaker   The loudspeakers used for pop music concerts are not suitable for this type of work. They often have distortion deliberately built in and tend to be too bassy. Ordinary hi-fi loudspeakers of sufficient power handling capability could be used, though many of these also tend to give an unnatural bass which some music-lovers may detect. A high-quality column fitted with adequately-sized drivers is probably the best, as the bass with these is more natural than with reflex or infinite baffle types. The column also has the advantage of having twice the range of other types so good coverage can be maintained right to the back rows. Given our English climate they should also be well waterproofed.

Canopy   There are various possible shapes and constructions for the platform canopy. One type sometimes seen is a half-cylinder made of corrugated metal similar to an aircraft hangar, with a closed back. The acoustic effect is that of an open pipe and so has strong resonances at its fundamental and odd harmonics, although these are damped to some extent by the presence of the orchestra within it.

An improvement is obtained by tapering the area toward the back, as this reduces the effect of harmonics, and projects the sound outward. One way of achieving a vertical taper is to arrange the orchestra in tiers. A better effect is obtained when both sides and roof diverge outward as it gives a horn-like configuration. If the internal contour is rounded to give a shape like a deep shell, this is the best of all. This may be difficult to construct, so a good compromise is to have the walls and back arranged in a curve, topped by a flat sloping roof. The roof should extend beyond the platform so as to minimise the amount of sound radiated upward, and reflect it downward toward the audience. The extension will also give added protection against rain.

The surfaces should be hard and reflective, brick or concrete blocks for the walls and solid or ply $^3/_4$ inch timber for the underside of the roof. Thinner ply on battens absorbs low frequencies so is not suitable. The walls can be faced with thin ply for decorative purposes providing it fixes directly to the stonework and is not mounted on battens. For temporary canopies corrugated steel is about the only practical material as it is reflective, fairly rigid and can be easily assembled on a scaffolding framework.

Most likely the sound engineer will not be consulted as to the platform and its construction and so will have to make the best of what he finds. If he does have a hand in a permanent installation perhaps he could arrange for a couple of large column loudspeakers to be built into a pair of wings on either side of the platform. The exterior finish could be such as to disguise their true nature. This would really be the ideal set up.

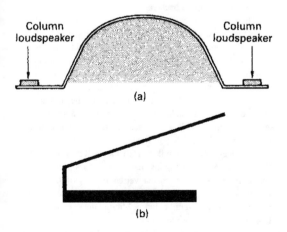

Figure 110. A practical and effective canopy construction: (a) top view; (b) side view.

Microphones

Studio capacitor microphones are preferred, but otherwise high-quality moving-coil units can be used.

There are two principal miking methods. One is to use only a single microphone, roof-mounted over the head of the conductor. The theory is that it will 'hear' what he hears; he is responsible for achieving the correct balance, and modifies the orchestra response to get the effect he wants. So that is what should go out to the audience: they came to hear the conductor's interpretation of the work, not the sound engineer's.

This is fine in theory if the acoustic of the canopy is ideal. If it is not, such as when a half-cylinder construction is employed, some instruments may sound hollow or boxy when picked up by a solitary microphone. A distant microphone, (as a single one must be from most instruments), is far more affected by environmental acoustics than one that is close. As high gain must be used there is also the danger of feedback.

Hence the other method is to use a number of microphones, which should be cardioid or hypercardioid so that they can be directed to specific instruments or sections. One objective is to avoid picking up brass instruments directly as they will swamp everything else; the directivity of the microphones will help do this if they are carefully angled. An exception is the french horns which have less output than the other brass instruments and direct their sound backward.

The woodwind section need special attention especially the double reed instruments such as the oboe, cor-anglais and bassoon. A single microphone each can be used to pick up the string sections of first and second violins, violas cellos and double basses, but care must be taken that it is roughly equidistant from all or most players in that section. If it is closer to one or two, these will dominate, and the homogeneous string effect which conductors try so hard to obtain will be lost. A microphone could be placed near the leader and kept turned down until he has a solo part to play.

A danger with close miking is that a small movement by the player, say by moving his chair a few inches or leaning forward or backward, or by inadvertent moving of the microphone stand, can have a large effect on the microphone output, hence the overall balance. Microphones should thus not be positioned too close to the player.

The control console and mixer should be out of sight of the audience as this would betray the use of the p.a. system. So what about the microphones on stage? With a large number of music stands about, these probably would not be noticed, especially at a distance. If stands and microphones are finished in black, they will be even less conspicuous.

A single boundary microphone could be an alternative to conventional units (see page 50).

Ambiophony If musicians cannot hear reverberation from their instruments, they tend to 'force' their tone which is obviously detrimental to their playing. This is likely when playing in the open, especially for the strings playing nearest the edge of the platform, but much depends on the acoustics of the canopy. Reverberation can be supplied artificially by means of a few small loudspeakers on stage fed from a reverb unit and small amplifier at low volume. This is called ambiophony.

Its need can be determined by standing on stage at various places and clapping the hands. A weak dry sound indicates lack of sufficient reverberation, whereas a full persisting sound shows it is adequate.

Induction loops

All NHS hearing aids issued since 1974 have a selector which enables the user to switch between microphone (M) and telephone (T). The latter switches in a coil which picks up the magnetic fields from some telephone earpieces and gives less distortion than the acoustic pickup between the microphone and earpiece.

When switched to T, the hearing aid can also pick up the field from an induction loop linked to a p.a. system. This gives a clearer sound, free from hall reverberation and ambient noise, which can be very confusing to hearing aid users.

Loop current The loop consists of several turns of wire run around the walls of the auditorium. The average field required inside a circular loop is obtained from a current of 100 mA per metre of loop diameter for a single turn. For a square loop the current is 112 mA times the length of one side. For a rectangular loop which is the most common configuration, it is 112 mA times the longest side, up to 1.5 times the width. For longer rectangles the figure remains at 1.5 times the width. These are for average signal levels, but allowance is needed for peaks up to 10 dB higher, so the available current must be multiplied by three. There is a further optimum height factor of 1.1. All this is divided by the number of turns, so total current required in amps is:

$$I = \frac{0.37 \times L}{t}$$

where L is the rectangle length up to 1.5 x width; t is number of turns.

Loop resistance   To keep the power requirement low, the loop resistance must also be low, thereby achieving a high current at low voltage. Around 5 Ω is the target which can be comfortably supplied from a 4 Ω amplifier output.

If the amplifier is feeding an existing load, the resistance must be higher than 5 Ω, being calculated not to overload the amplifier when shunted across its present load. Usually the loop requires more power than the loudspeakers, so its resistance should be lower than the equivalent impedance of the speaker load. If this is not possible without overloading the amplifier, a separate amplifier should be used.

When the target resistance is determined and the loop path measured to discover the total length, the cable resistance table can be consulted to select the gauge which gives the closest resistance to the target. Three turns is the most convenient number because it can be formed from a single three-core mains cable having the cores connected in series. For small halls four turns made up of two twin cables may be needed to make up the required resistance.

Loop height   Maximum pickup occurs when the pickup coil has the same orientation as the loop, that is both with vertical axes, and when they are in the same plane. However, in this position the field across the loop is strong at the sides and reduced at the centre, giving uneven results. The effect can be compensated for by raising or lowering the loop to a different plane from the pickup coil. The circular field that surrounds the loop, then, is little changed in orientation at the centre, but progressively changes toward the sides. This evens out the signal pickup, the optimum being when the vertical displacement between the loop and the pickup coil is 0.1 times the width of the loop.

Power requirement   Power required is:

$$W = \left(\frac{0.37 \times l}{t}\right)^2 R$$

where W is the power in watts; l is the length of the loop up to 1.5 times the width; t is the number of turns; and R is the resistance.

With the loop displaced vertically to the optimum 0.1 width dimension, the current required is 1.1 times, and the power 1.2 times, that needed when the loop and pickup coils are in the same plane. The formula includes this factor. At larger displacements the current and power requirements are considerably greater.

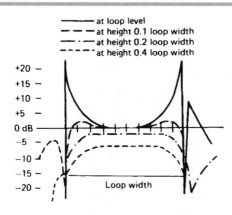

Figure 111. Signal pickup at various vertical displacements across the loop. Horizontal calibration in tenths of loop width. Overspill shown on one side only to avoid confusion. Optimum is at 0.1 loop-width displacement.

Wire resistance table

Gauge	Ω/core/100 m	Gauge	Ω/core/100 m
1/0.1 mm	219	24/0.2 mm	2.4
1/0.2 mm	57.6	32/0.2 mm	1.78
1/0.6 mm	6.4	40/0.2 mm	1.44
1/0.8 mm	3.6	30/0.25 mm	1.17
7/0.2 mm	8.2	1.0 mm²	1.8
13/0.2 mm	4.4	1.5 mm²	1.2
16/0.2 mm	3.6	2.5 mm²	0.72

Current and power increases with vertical displacement

Width/height ratio	Current x	Power x
0.1	1.1	1.2
0.2	1.25	1.6
0.3	1.5	2.25
0.4	2.0	4.0
0.5	2.5	6.25
0.6	3.25	10.6
0.7	4.25	18.0
0.8	5.5	30.2
0.9	7.0	49.0
1.0	8.5	72.2

P.a. system problems

Instability This is caused by unwanted coupling between two points between which there is amplification. There are three types, capacitive, magnetic, and common impedance. At audio frequencies capacitive coupling is rarely a problem, but magnetic coupling between speaker feeders and microphone cables are possible. A high-pitched whistle is the usual symptom, but it can be supersonic and although not audible, result in amplifier overloading and distortion. It can be identified by a steady reading on the output meters with no signal, and is avoided by keeping speaker and microphone cables well apart, especially over long runs.

Common impedance coupling occurs when an impedance is common to both input and output circuits. A common example is when the non-earthed side of a speaker feeder develops a leak to earth. There is then a path from one side of the output transformer secondary to earth, and from the other side through the amplifier, and the screened lead to the mixer, to earth. Part of the output signal is thus travelling along the braid of the mixer lead together with the input signal to the amplifier, and coupling takes place. With all large installations a routine check for earth leakage from the speaker distribution circuit should be made. Common impedance coupling can also be caused in the amplifier or mixer by the failure of a decoupling capacitor.

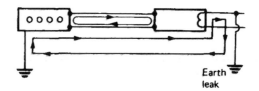

Figure 112. Mixer lead screen forms a common impedance for input and output signal when a speaker feeder has a leak to earth.

Earth loop If more than one item of equipment is earthed, an earth loop can be formed. Any hum fields through which a screened cable may pass, induce a voltage in the braid. If there is an earth at both ends, a complete circuit exists which permits a hum current to flow. As the screen is part of an input circuit, the current is effectively in series with the input signal, and hum results.

The standard method of preventing an earth loop in the past has been to earth the system at only one point, preferably to the mixer, with auxiliary items such as tape recorders and grams being isolated from earth. Recent regulations have barred this (see the next section) as all units in the system must now be earthed.

The alternative is to open-circuit the braid at one end of all screened connecting leads. The braid remains earthed at the connected end but does not offer a continuous path for hum currents. The signal return paths is now via the respective earths.

As these may themselves pickup hum voltages, a further precaution is to bond all the unit cases together with an earth wire. Units mounted in steel racks will be immune from this problem as the cases are all connected via the rack.

A loop can be formed when tape-in and tape-out leads are taken from a tape recorder to a mixer to afford record and play facilities. In this case the braid of one lead only should be open-circuited.

Radio interference   Breakthrough from regular broadcast and mobile radio transmitters is often experienced with p.a. systems and domestic hi-fi equipment. As radio sources proliferate it has become a major problem in some areas, with bursts of speech or continuous background music intruding on church services and meetings.

The interfering signal is picked up on the screen of one of the input leads, which is particularly vulnerable if it is a quarter of the signal wavelength long or an exact multiple. As the screen is in series with the source, the signal is presented to the first stage, which having no radio tuned circuits, accepts all radio frequencies received. The rectifying action of the base/emitter junction of a bipolar transistor demodulates the interference which is then amplified along with the regular signal. Field-effect transistors do not have a rectifying action, and so are often used in input circuits to avoid radio breakthrough. Balanced microphone cables are less prone to the problem because the screen is no part of the input circuit, but unbalanced ones are susceptible. Even when radio interference suppression is provided in the mixer, it can breakthrough if a microphone socket is tarnished or oxidised. A rectifying junction can thereby be formed which demodulates the radio signal before it reaches the suppression circuits. All microphone contacts should therefore be kept bright and clean.

Suppression   In many cases a capacitor connected across the mixer input sockets will bypass radio interference. The value depends on the input impedance. The reactance must be low in comparison to the load at the radio frequencies it must suppress, but high in comparison to the load at the highest audio frequencies. As most mobile transmitters operate at vhf, there is a wide gap between them and the audio band, so a simple capacitor bypass of 0.015 μF should suffice, but it should have low self-inductance as this would offer an impedance at radio frequencies and so reduce the effectiveness of the bypass.

The problem often arises with long microphone cables. Sometimes one particular cable and input will be more prone to interference than others. If this is so it could be that the cable is at a critical length which is a multiple of the interfering wavelength. Shortening the cable by about a foot may help. Alternatively it could be that the plug/socket connections are tarnished and rectifying received radio signals before reaching the mixer.

If the bypass capacitor at the mixer input socket fails to cure the trouble better results may be obtained by fitting an additional one at the platform socket. As the values are then added, they must be less than that of the single capacitor if the upper audio frequencies are not to be curtailed; 0.01 μF at each end should be suitable. The inductance of the microphone cable then forms a pi filter with the capacitors.

With some public address installations microphone cables are run over a suspended ceiling from sockets on the platform to a control desk elsewhere in the hall. Although a convenient way to install, this is a bad practice as the elevation makes them very prone to radio interference. Ideally, microphone cables should be run below the floor. If a new building is being constructed, conduit can be laid in the concrete floor-bed to receive the cables later. Above ceiling runs should only be attempted if the cables are run in earthed copper tubing.

A further consideration is whether it is necessary to have the control desk at a great distance from the platform. A near location means short microphone cables and reduced probability of radio interference. It has other advantages too, such as a clear view of the platform for the operator, and near access for him to correct any problems with microphones that may arise.

Sometimes interference is experienced from powerful nearby medium or long-wave transmitters. A filter with a sharper cut-off may then be needed; with a series coil being added to the circuit. If there is one particular transmitter causing the trouble, a wave-trap consisting of a series coil and capacitor tuned to resonate at the transmitter frequency, should be shunted across the input. Otherwise a general purpose filter should be tried.

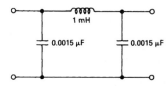

Figure 113. Circuit of a pi microphone radio interference input filter.

Electrical interference   Crackles and pops generated by nearby thermostats and other equipment can sometimes be troublesome. Large transient surges are produced which contain high radio frequencies, so these affect the equipment in a similar way to radio interference. There are two ways whereby they can gain access. One is by radiation picked up by the input leads, and the other is via the mains wiring. The first can be dealt with by radio suppression, and the latter by filter capacitors across the mains input to the equipment. Suppression units are available which can conveniently be plugged into a mains socket.

Cables   Many p.a. system faults are due to damaged cables. Microphone leads are particularly vulnerable as they experience much flexing. Trouble can be minimised if good quality padded leads are used with rubber or spring strain relievers at the ends. Most breaks occur within inches of the ends, and the most practical method of dealing with them is to cut off two or three inches and remake the connection into the plug. If it is not known which end is at fault, it can be determined by connecting a capacitance meter or bridge across the lead at each end in turn. The one giving a low capacitance reading is the one with the break.

Breaks or shorts are sometimes encountered with long speaker feeders. The location of breaks can be found by measuring the capacitance and comparing with that of a known length of cable; while shorts can be found by comparing resistance readings with the wire table.

Public-address safety regulations

A number of regulations have recently been introduced to supplement the Health and Safety at Work Act (1974). These are contained in the Electricity at Work Act (1990), and the guidelines laid down by the Health and Safety Executive (GS 50 1991). Licences for public gatherings can be refused by local health and safety inspectors, and their interpretations can vary.

The new rules concern mains supplies, earthing, 100 V feeders, and cable runs. Some are sensible, many are not, but unfortunately all have to be complied with, though causing unnecessary difficulty for the installer.

Mains supplies   All supplies to any system must be from the same mains phase to avoid 400 V appearing across insulation intended only for 240 V. The only problem here could be a remote mains-powered mixer in a large outdoor installation.

Residual current devices (RCDs) rated to trip at 30 mA should be employed, each supplying no more than six outlets. Ideally they should be mounted on the sub-circuit distribution board. RCD adapters or plugs should not be used as these can be easily removed. For permanent systems they must be tested by operating the trip button once a month, but for temporary installations they must be tested each day.

Excess current protection must be afforded either by correctly rated fuses, or by circuit breakers (CBs). These may be combined with residual current devices (RCBOs)

Earth connections   All items of equipment must be separately earthed either via individual mains cable earth wires or by a bonding wire connecting the cases of all units, or by mounting all in a steel rack. The practice of earthing only the mixer and removing all other earths to avoid hum loops is no longer acceptable. Loops can be avoided by using a screened connecting cable in which the braiding is open-circuited at one end. An item need not be earthed if it is double insulated and bears the twin concentric rectangle mark.

100 V lines   It is now stipulated by the International Electro Technical Commission that an a.c. voltage over 50 V is dangerous and potentially lethal; 25 V is thus deemed the maximum safe voltage and 100 V lines are considered to be dangerous.

This ignores the fact that a 100 V continuous sine wave is never encountered. It can only be present when an amplifier is run at maximum output, but as clipping starts at that point the universal practice is to operate with peaks at least 3 dB below full output, which gives a maximum of 70 V. The average level is some 10 dB lower than that, giving a signal output voltage of around 24 V. Even this is not continuous as there are many breaks in a normal speech signal.

The danger is thus considerably overrated if any exists at all. The 100 V designation is really only a technical convention to describe the output when an amplifier is being tested with a sine wave at full output. This is a classic example of what happens when bureaucrats legislate concerning matters they know nothing about!

Cables　The practical effect for the installer is that double-insulated cable must be used, and tap-in connections as commonly used with temporary systems are not allowed. Thus all connecting cables must be made off with suitable connectors at their ends. This will cause many headaches at large temporary installations where ready-made leads will always be of the wrong length.

With permanent systems there is no problem as permanent wiring and connects will be made. In permanent large outdoor systems such as at sports stadiums, the cable must be run in conduit to protect it from damage. The reason for this is that the p.a. system may be required to broadcast life-saving instructions in the event of a major disaster such as a fire.

A permanently installed system, used primarily for spot announcements may not be of a sufficiently high standard for continuous use such as meetings and rallies. A superior temporary system may therefore be installed, but as the cable for this cannot practically be protected in conduit, the permanent system must always be on standby in case of an emergency.

All temporary cables must be run so that there is no danger from tripping or overhead hazard. A 10 ft minimum height is stipulated, increasing to 18 ft where vehicles may pass. Catenary wires should be used for long spans where the weight of the cable may approach its strain strength.

Mixer and amplifiers　There must be no exposed terminals bearing more than 25 V, or if higher they should deliver no more than 5 mA when short-circuited. Resistance from equipment cases to earth pins must be less than 0.1 Ω or for items fused at less than 3 A, 0.5 Ω All equipment insulation should be tested at 500 V a.c. for more than 5 seconds. These tests must be carried out regularly by a competent person who should record the date and the readings obtained. In addition visual checks on all mains cables and plugs, including a pulling and twisting test to ensure secure cable clamping, should be made and the results recorded.

Facts and formulae

Resistance

Ohm's law: $\quad V = IR; \quad I = \dfrac{V}{R}; \quad W = I^2R; \quad W = \dfrac{V^2}{R}$

in which V = volts; I = amperes; R = ohms; W = watts.

Resistance in series: $\qquad R = r_1 + r_2 + r_3 \ldots$

Resistance in parallel:

$$R = \frac{r_1 \times r_2}{r_1 + r_2} \quad \text{or} \quad \frac{1}{R} = \frac{1}{r_1} + \frac{1}{r_2} + \frac{1}{r_3} \ldots$$

Reactance

$$X_L = 2\pi fL \qquad X_C = \frac{1}{2\pi fC}$$

in which L = henries; C = farads

Impedance

Series capacitance/resistance: $\qquad Z = \sqrt{R^2 + X_C^{\,2}}$

phase angle: $\qquad \phi = \tan^{-1} \dfrac{X_C}{R}$

Series inductance/resistance: $\qquad Z = \sqrt{R^2 + X_L^{\,2}}$

phase angle: $\qquad \phi = \tan^{-1} \dfrac{X_L}{R}$

Series inductance/capacitance/resistance:

$$Z = \sqrt{R^2 + (X_L - X_C)^2}$$

phase angle: $\qquad \phi = \tan^{-1} \dfrac{(X_L - X_C)}{R}$

Parallel capacitance/resistance: $\qquad Z = \dfrac{R \times X_C}{\sqrt{R^2 + X_C^{\,2}}}$

phase angle: $\qquad \phi = \tan^{-1} \dfrac{R}{X_C}$

Parallel inductance/resistance:
$$Z = \frac{R \times X_L}{\sqrt{R^2 + X_L^2}}$$

phase angle:
$$\phi = \tan^{-1} \frac{R}{X_L}$$

Series inductance/resistance, parallel capacitance:

$$Z = \frac{1}{\sqrt{\left(\dfrac{R}{R^2 + X_L^2}\right)^2 + \left(\dfrac{X_L}{R^2 + X_L^2} - X_C\right)^2}}$$

phase angle:
$$\phi = \tan^{-1} \frac{\left(\dfrac{X_L}{R^2 + X_L^2} - X_C\right)}{\left(\dfrac{R}{R^2 + X_L^2}\right)}$$

Resonance (series)

When inductive and capacitive reactances are equal at f_r, the resonant frequency; hence:

$$X_L = X_C; \quad f_r = \frac{1}{2\pi\sqrt{LC}}$$

Impedance at resonance: $Z = R$

Resonance (parallel)

When there is zero phase difference and reactance, reactances are not exactly equal so:

$$f_r = \frac{1}{2\pi}\sqrt{\frac{1}{LC} - \frac{r^2}{L^2}}$$

Impedance at resonance: $Z = \dfrac{L}{CR}$

Adding two AC voltages or currents

$$V = \sqrt{\cos\theta \times 2v_1 \times v_2 + v_1^2 + v_2^2}$$

in which θ is the angle of phase difference; v_1 is the first voltage or current; v_2 is the second voltage or current.

Q-factor

Q defines the quality of a component or circuit as:

$$\frac{2\pi \times \text{maximum energy stored in 1 cycle}}{\text{energy dissipated in 1 cycle}}$$

Inductor: energy stored = LI^2 joules; energy dissipated :

$$\frac{I^2 R}{f} \text{ joules,}$$

hence: $Q = \dfrac{2\pi L I^2 f}{I^2 R}$ or $\dfrac{2\pi f L}{R}$ which is $\dfrac{X_L}{R}$

Q thus increases with frequency until *skin effect* increases R, causing Q to rise less rapidly. Then the self-capacitance reduces X_L and also Q, until it is completely cancelled and Q falls to zero. This occurs at the self-resonant frequency of the inductor.

Capacitor: $Q = \dfrac{1}{2\pi f C R}$ or $\dfrac{1}{X_C R}$

Resonance (series circuit): $Q = \dfrac{1}{R}\sqrt{\dfrac{L}{C}}$

The voltage across either the capacitor or inductor at resonance is Q times the applied voltage, hence Q is also the *magnification factor*.

Resonance (parallel circuit): assuming negligible loss in the capacitor, the Q is equal to that of the inductor. With a Q higher than 10, the current in each branch is Q times the supply current, so Q is a current magnification factor.

Bandwidth: bandwidth at the -3 dB points either side of the resonant frequency is inversely proportional to Q and is given by:

$$\frac{f_r}{Q}$$

Skin effect   When an alternating current flows along a conductor its central area intercepts magnetic flux generated by all regions, but at the surface area much of the flux extends outside the conductor. The field is thus more concentrated at the centre, and the back e.m.f. it generates is stronger there. As this opposes the original current flow, more resistance is offered than in the surface regions. Put another way, the inductance is greater at the centre. The overall resistance is thereby increased with most of the current flowing on the surface.

The effect is expressed as *Rac/Rdc*, the ratio of a.c. resistance to d.c. It rises with frequency and conductor diameter. The relationship is a complex one, but an approximate value can be obtained from:

$$Rac \:/\: Rdc \approx 4d\sqrt{f} \quad \text{hence} \quad Rac \approx 4d\sqrt{f}\,Rdc$$

in which *d* is the diameter of the conductor in mm; and *f* is the frequency in MHz. The ratio becomes much less accurate below 200 kHz where it is very small.

Time constant   Defined as the time in seconds taken for the voltage across a capacitor to reach 63% of its final value through a series resistance, also the time in seconds taken for the current through an inductor to reach 63% of its final value through a series resistance:

$$t = CR \qquad t = \frac{L}{R}$$

Attenuators and matching pads

Resistor networks are used for attenuation or for matching a circuit of one impedance to that of another. The main types are: L, T, and π, so called because the circuit configuration resembles those letters. Each type except L can be *symmetrical*, *unsymmetrical*, *balanced* or *unbalanced*.

L-type network   This is used for matching two dissimilar impedances with the minimum attenuation. The high (Z_H) must be connected to terminals 1 and 2, and the low (Z_L) to 3 and 4, irrespective of which is the source and which is the load (see page 297). The values are calculated by:

$$R_1 = \frac{Z_H \times Z_L}{R_2} \qquad R_2 = \sqrt{\frac{Z_H \times Z_L^{\,2}}{Z_H - Z_L}}$$

The insertion loss in dB is:

$$10\log_{10}\left(\frac{Z_L + R_2}{R_2}\right)^2 \times \frac{Z_H}{Z_L}$$

Balanced L-network   A balanced version of the L-network can be produced by splitting the value of R_1 into two equal parts and putting them in each series leg.

T-type network   The symmetrical T-type pad is used between two circuits of the same impedance (Z) to introduce attenuation without upsetting the existing match. The design is simplified if the required attenuation is defined as a current ratio (N) rather than in decibels. Then:

$$R_1 = R_2 = \left(\frac{N-1}{N+1}\right) \qquad R_3 = Z\left(\frac{2N}{N^2-1}\right)$$

Asymmetrical T-type network   The asymmetrical T-pad is used to insert a required amount of attenuation (N) between a high impedance (Z_H) and a low impedance (Z_L) circuit. There is a minimum loss, so attenuation cannot be lower than this. It is governed by the ratio of the two impedances, so $N > Z_H/Z_L$. The resistor values are:

$$R_3 = \sqrt{Z_H Z_L}\left(\frac{2N}{N^2-1}\right) \qquad R_1 = Z_H\left(\frac{N^2+1}{N^2-1}\right) - R_3$$

$$R_2 = Z_L\left(\frac{N^2+1}{N^2-1}\right) - R_3$$

The high impedance (Z_H) must be connected across terminals 1 and 2, while the low impedance (Z_L) is connected across terminals 3 and 4.

Balanced T-network   The network can be balanced by halving the values of R_1 and R_2, and adding R_4 and R_5 with values $R_1 = R_4$, and $R_2 = R_5$.

Bridged T-network   This pad is used for switched attenuators where several values of attenuation are required. Although an extra resistor is used for bridging each network, two are common to all, so for a three-level attenuator eight resistors are used instead of nine. By bridging R_1 and R_2 of a symmetrical T-pad with R_4, the attenuation can be changed by altering only two resistors, R_3 and R_4, instead of three. Thus the switch need have only two poles. The values are:

$$R_1 = R_2 = Z \qquad R_3 = \frac{Z}{N-1} \qquad R_4 = Z(N-1)$$

π-type network   A π-type pad has the same properties as the T-pad and can be used as an alternative, having the same number of resistors. In some cases the values of one type may be closer to preferred values than of the others. The shunt resistance of both types are the same, but the series value is higher with the π-type, which may be a consideration if d.c. is involved. Values for the symmetrical type are:

$$R_1 = R_3 = Z\left(\frac{N+1}{N-1}\right) \qquad R_2 = Z\left(\frac{N^2-1}{2N}\right)$$

Asymmetrical π-type network Values for this are more complex and are most easily calculated by first designing an asymmetrical T-pad then using its values for the π-pad as follows:

$$R_1 = \frac{R}{TR_2} \qquad R_2 = \frac{R}{TR_3} \qquad R_3 = \frac{R}{TR_1}$$

in which *TR* are the T-pad values; and
$$R = TR_1TR_2 + TR_1TR_3 + TR_2TR_3$$

Balanced π-type network Balancing is achieved by halving the value of R_2 and adding R_4 of the same value. Thus only four resistors are needed instead of five for the equivalent balanced T-type network.

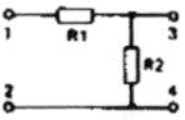

(a) unbalanced L-type network 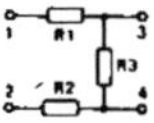   (b) balanced L-type network

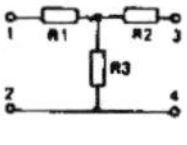

(c) unbalanced T-type network 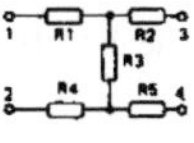   (d) balanced T-type network

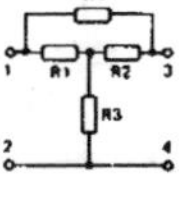

(e) bridged T-network 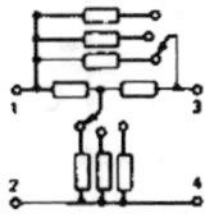   (f) 3-level attenuator using bridged T-pads

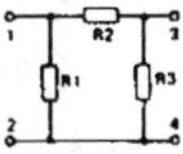

(g) unbalanced π-type network 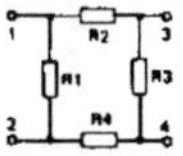   (h) balanced π-type network

Figure 114. Various types of attenuator, balanced and unbalanced.

Filters

Impedance The behaviour of a filter is determined by the terminating impedance, so it is designed for a particular value which is termed the *design impedance*. When so designed, the impedance measured at the input terminals is the same as that of the output and is called the *iterative impedance*. Filters may then be cascaded providing the last one is terminated with the correct impedance.

Constant-k low-pass filters These can be either T-type or π-type having a nominal 18 dB/octave attenuation above the cut-off frequency which is –3 dB below the pass-band level. Both types have the same properties, but as the π-type has only one inductor against the two of the T-type, it is generally preferred. With the T-type the calculated inductance is halved for each of the values of the two inductors and the capacitance value is halved for the two capacitors of the π-type to give the values of the two capacitors. The same formulae apply to both types:

$$Z = \sqrt{\frac{L}{C}} \qquad f = \frac{1}{\pi\sqrt{LC}} \qquad L = \frac{Z}{\pi f} \qquad C = \frac{1}{\pi f Z}$$

in which L is the inductance in henries; C is the capacitance in farads; f is the cut-off frequency; and Z is the design impedance.

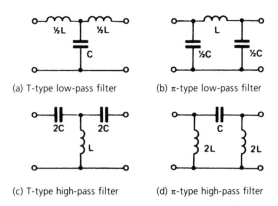

(a) T-type low-pass filter   (b) π-type low-pass filter

(c) T-type high-pass filter   (d) π-type high-pass filter

Figure 115. Various constant-k filters.

Constant-k high-pass filters Either T-type or π-type filters can be used to provide the nominal 18 dB/octave attenuation below the cut-off frequency. Both have the same properties, but here the T-type has only one inductor whereas the π-type has two. Choice may be influenced by the fact that the π-type has a d.c. path across input and output whereas the T-type has not. The calculated capacitance is doubled for each of the capacitors of the T-type filter, while the inductance is doubled for each inductor of the π-type. The formulae for both is:

$$Z = \sqrt{\frac{L}{C}} \qquad f = \frac{1}{4\pi\sqrt{LC}} \qquad L = \frac{Z}{4\pi f} \qquad C = \frac{1}{4\pi f Z}$$

M-derived filters These include a tuned circuit which gives a very high attenuation at the resonant frequency, but a much lower attenuation than a constant-k filter beyond it. M is a function of the ratio between the cut-off frequency f_c and the resonant frequency f_r. It is:

$$m = \sqrt{1 - \left(\frac{f_c}{f_r}\right)^2}$$

An m value of 1.0 is equivalent to a constant-k filter. As it gets lower, attenuation at the cut-off frequency gets steeper, but that above it for low-pass filters, and below it for high-pass filters, becomes less.

Attenuation beyond cut-off drops to the levels below

M values	0.4	0.5	0.6	0.7	0.8	0.9
dB attenuation	7.6	10	12	15	19	26

T-type, M-derived low-pass filter Circuit is the same as for a constant-k filter, but a third inductor is added in series with the capacitor. The values are calculated for constant-k then related to m as follows:

$$L = \frac{Z}{\pi f} \qquad C = \frac{1}{\pi f Z} \qquad l_1 = l_2 = \frac{mL}{2} \qquad l_3 = \frac{(1-m^2)}{4m}L \qquad c = mC$$

π-type, M-derived low-pass filter   A third capacitor is shunted across the series inductor of the constant-k circuit. Having one inductor to the T-type's three makes it an obvious choice. There is no d.c. path in either. Values are derived from the constant-k formulae:

$$L = \frac{Z}{\pi f} \quad C = \frac{1}{\pi f Z} \quad c_1 = c_2 = \frac{mC}{2} \quad c_3 = \frac{(1-m^2)}{4m}C \quad l = mL$$

T-type, M-derived high-pass filter   A third capacitor is added in series with the inductor of the constant-k circuit. M modified values are:

$$L = \frac{Z}{4\pi f} \quad C = \frac{1}{4\pi f Z} \quad c_1 = c_2 = \frac{2C}{m} \quad c_3 = \frac{4m}{1-m^2}C \quad l = \frac{L}{m}$$

π-type, M-derived high-pass filter   The series capacitor in the equivalent constant-k circuit is shunted by a third inductor. The values are:

$$L = \frac{Z}{4\pi f} \quad C = \frac{1}{4\pi f Z} \quad l_1 = l_2 = \frac{2L}{m} \quad l_3 = \frac{4m}{1-m^2}L \quad c = C$$

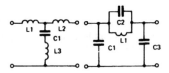

(a) T-type, m-derived low-pass   (b) π-type, m-derived low-pass

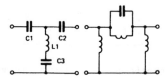

(c) T-type, m-derived high-pass   (d) π-type, m-derived high-pass

Figure 116. Various m-derived filters.

Test equipment

A well-equipped workshop is essential to the professional audio engineer, and desirable for the amateur. Some items are vital, others useful, while yet others are desirable but dispensable. Much depends on the type of work undertaken. We here briefly survey the main items and their uses.

Multi-range test meter   This is the first essential for checking a.c. or d.c. voltages, current and resistance. For audio work an analogue meter is preferable to a digital one because it can give an intelligible indication of rapidly changing readings such as when used as an output meter, whereas a digital unit gives a meaningless jumble of figures. Extensive voltage ranges are not essential, but high current a.c. and d.c. ranges are. The meter should have a clear readable scale, not too cramped on the resistance ranges, and be capable of standing hard knocks.

Millivoltmeter   Few multi-range meters read millivolts, so a separate meter is needed as many tests involve measurements in this region. One may be included in audio analysers or test sets.

Oscilloscope   Another essential item, the oscilloscope reveals clipping, ringing, distortion, phase displacement, noise, hum, and other ills. The wideband X amplifiers found in models used for TV servicing are not necessary for audio work, but a sensitive Y amplifier is essential for examining small signals. A double beam facility can be useful to compare two signals such as an input and output. Avoid too small a screen with which small details could be missed (see next section).

Audio generator   A generator covering the audio range is a further essential. If distortion measurements are contemplated the generator should itself produce very low distortion, because its level sets the lower limit of what can be measured. Frequency calibration needs to be moderately accurate but not excessively so. More important is accurate output level calibration or indication. Frequency response and sensitivity tests require audio signals of accurately known levels. Some generators do not maintain the output calibration accuracy over the whole frequency range. This is permissible provided the instrument has an output meter which enables required levels to be set. The generator should have a wide output range from below 1 mV up to 10 V. It should also have a square-wave output as this is invaluable for quick frequency response checks, tone control calibration, and ringing tests (see next section).

Pink noise generator   White noise consists of random frequencies spread fairly evenly over the audio spectrum. As each octave has a frequency range equal to all those below it, most of the energy of white noise is concentrated in the upper octaves. With pink noise, the energy is divided equally between the octaves. This corresponds to the human hearing characteristic and so makes an appropriate test signal.

It is particularly useful for testing loudspeaker installations, either domestic or public address; levels can be checked with a sound meter in all parts of a room or auditorium. Sensitivities of loudspeakers and microphones can be compared, and microphone sensitivities equalised for making feedback tests. If these are attempted with pure tones from an audio generator, sound reflections cause cancellations and reinforcements giving misleading results. Pink noise consisting of many simultaneous frequencies swamp interference effects at individual frequencies. Recorded pink noise can make a suitable though less convenient substitute for a generator.

Sound meter   This can be anything from a simple device measuring sound according to the A-weighting curve, to complex instruments with various switchable curves, filters, impulse measuring, averaging and peak-hold facilities. Usually the microphone is mounted on an extension arm to minimise diffraction effects in an omnidirectional field, but not with some of the cheaper ones. The full specification models are used by those working with architectural acoustics and noise measurement, but for general sound level checking and p.a. work, the simpler models are quite adequate. They are reasonably priced and well worth including in the test kit.

Output meter   A meter measuring watts but not usually a wattmeter with voltage and current coils. It consists of a voltmeter calibrated to read watts with a given load resistance. Different loads can be switched in and series meter resistors switched at the same time to maintain the correct reading. Useful for amplifier output testing, but the multi-range meter can do the same thing with an external load resistor, The watts must be calculated from the voltage reading by E^2/R.

Harmonic distortion meter   This consists basically of a voltmeter calibrated in percentages and a tunable notch filter. A pure tone from a signal generator is fed into the equipment under test which is properly terminated. The distortion meter is connected to the output and the levels set to give 100%

reading. Then the filter is switched in and tuned to give minimum output, switching down the ranges to maintain a readable indication. The fundamental is thus eliminated and the residue which comprises the spurious harmonics is measured as a percentage of the whole. The reading is not quite accurate as it includes noise. Distortion measurements are not frequently required in a repair workshop. but are essential for production testing, quality control and equipment reviewing.

Intermodulation distortion meter   Two frequencies, a high and a low, are generated, not necessarily of low distortion, and are applied in a fixed proportion to the equipment under test. The output is passed through a high-pass filter to remove the low frequency, and the remaining high is demodulated. Any resulting signal is an intermodulation product between the two frequencies and is measured as a percentage of the original. Like the harmonic distortion meter its use is limited to specific applications.

Transistor tester   Most equipment faults are due to semiconductor failures. Although i.c.s are being increasingly used, discrete transistors are still common. A tester, reading h_{fe}, leakage and resistance is very useful for fault checking and matching up complementary pairs.

Capacitance meter   A capacitance bridge was once the workshop norm, but inexpensive, accurate, direct reading instruments that range from a few pF to several thousand μF are now available and are far more convenient to use. Next to transistors, capacitors are the most common causes of equipment failures. A meter can quickly test these and is also useful for locating cable breaks.

Inductance bridge   One of those instruments that gathers dust most of the time but worth its weight in gold when you are checking or making up filters and crossover networks and want to discover a coil's inductance.

Spectrum analyser   Audio signals are split into one-third octave bands and measured, so identifying circuit or acoustic resonances and the harmonic content of noise or distortion. Invaluable for much design and research and fascinating to experiment with, but beyond the scope and budgets of most workshops.

Oscilloscope traces

Many faults and conditions can be identified by judicious use of the oscilloscope. Apart from a multimeter it is the number one tool of the audio engineer. The following are some of its uses.

Overloading   Overloading occurs if the input signal is too high or some fault reduces the signal handling capability of any stage. It can be identified by the clipping off of the tops of the peaks to a straight line. All peaks of the same polarity clip at the same level and reducing the gain to below that level restores the original waveform, although distortion just below it will be high. By moving the oscilloscope input lead to various stages, the one responsible can be revealed.

Symmetrical clipping of both half-cycles is usual; if one clips before the other, there is an imbalance in a complementary stage most likely the output stage. Power output can be measured by increasing the input signal from an audio generator with the output meter, oscilloscope and dummy load connected, until clipping commences. Ease the gain back to just before the clipping level and the reading gives the maximum power output. The level of the input signal then also gives the sensitivity.

Harmonic distortion   Low distortion levels are not visible on a displayed sine wave, but larger ones associated with fault conditions are. The appearance of the distorted wave depends on the number of the harmonic, its amplitude, and the phase relationship to the fundamental. It is thus not easy to identify the nature of the distortion by its appearance, but there is one useful rule. If the distortion is *symmetrical*, it results from mostly odd harmonics, whereas if it is *asymmetrical*, it is mainly the result of even ones. As harmonics usually decrease in amplitude with order, it can be assumed that the third and second harmonics predominate in the above two cases. Some faults produce odd while others generate even harmonics, so this identification can be very helpful in diagnosis.

Phase difference   Phase differences between two channels or between an input and output can be observed by feeding in a sine wave and applying the output of one channel to the vertical, and that of the other to the horizontal oscilloscope input, with the internal time base switched off. When the outputs are in phase, the result is a straight double line tilted at

45° with the left side down. As the phase difference between the channels increases, the lines part to form an ellipse until a full circle is formed at 90°. Beyond this the circle narrows to an ellipse tilted in the opposite direction until it becomes a double line again at 180°. The phase angle can be determined by measuring the length of the ellipse E_L, and the width measured horizontally across its centre E_W, then:

$$\sin\phi = \frac{E_W}{E_L}$$

The set-up can also be used to check the phase of filter circuits, provided they are terminated with the correct impedance, and to adjust the head azimuth of stereo tape recorders as phase errors are introduced by an incorrect azimuth.

Lissajous figures   An unknown frequency can be determined by feeding it to one oscilloscope input and the signal from an audio generator to the other. The generator frequency is adjusted until a circle is obtained, when the two frequencies are equal. If a double or triple loop appears, one signal is twice or three times the frequency of the other.

Frequency response   A quick check is to inject a square wave and observe the result on the oscilloscope. Mathematically, a square wave consists of a sine wave fundamental and an infinite number of odd harmonics, so the shape of the trace reveals the proportion of harmonics.

The square-wave frequency is around 1 kHz which is approximately the halfway point of the audio spectrum. A perfect square wave trace indicates a flat and extended response. The rising edge is governed by the high frequencies and the trailing edge by the low.

A *rounded leading edge* reveals a poor high frequency response. The degree of rounding indicates the extent. A *heightened leading edge* giving a tilted top appears when there is an excessive high frequency response. A *rounded trailing edge* indicates a poor bass response and the amount again reveals the extent of the problem. A *raised trailing* edge with a tilted back top shows excessive low frequency response.

This affords a way of checking the 'flat' positions of tone controls which are not always as calibrated owing to component tolerances.

Stability check   Some amplifiers are unstable with capacitive loads. With no signal input the oscilloscope is connected to the output and various values of capacitance from 0.01–0.1 μF are shunted across it. No spurious signal should appear. The square wave affords another check. Any waviness along the top reveals incipient instability.

Tone-burst testing   A fixed tone is chopped into bursts so that a specified number of cycles are followed by a gap of similar length. This signal is fed to a loudspeaker under test in an anechoic environment and picked up by a test microphone. The result, displayed on a double-beam oscilloscope, is compared to the original signal. Ringing and overhang are displayed as a decaying train of oscillation after each burst. The test is especially revealing at the crossover frequencies.

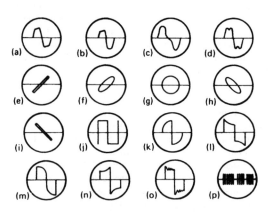

Figure 117. Clipping: (a) symmetrical; (b) asymmetrical. Distortion: (c) even harmonic; (d) odd harmonic. Phase traces: (e) in-phase; 45°; (g) 90°; (h) 135°; (i) 180° phase difference. Square waves: (j) ideal; (k) poor h.f.; (l) excessive h.f.; (m) poor l.f.; (n) excessive l.f.; (o) instability; (p) tone-burst.

Power circuits

The power supply circuit has many forms, but those used for audio applications comprise just a few basic types consisting of three essential components: the mains transformer, the rectifier and the reservoir capacitor.

Transformer   The primary is designed for 240 V mains voltage and the secondary for the required voltage, the turns ratio being proportional to the voltage ratio. For full-wave circuits and split-polarity bridge circuits, the secondary is centre-tapped. A transformer is rated in VA, which is the product of the secondary volts and amps, and has a regulation rating which specifies the percentage difference between the off-load voltage (V_{OL})and the full-load voltage (V_{FL}). The formula is:

$$\text{Regulation} = \frac{V_{OL} - V_{FL}}{V_{OL}} \times 100\%$$

Toroidal transformers have very low flux leakage, hence a minimum hum field, and so are preferred for audio equipment.

Rectifier   Silicon rectifiers are now universally used because of their small size and high current ratings. Voltage drop at low current is 0.6 V. When non-conducting, the combined transformer and reservoir capacitor voltage are effectively in series, and produce a peak inverse voltage of up to three times the applied r.m.s. a.c. voltage. This must be considered when replacing a rectifier.

Reservoir capacitor   This charges from the applied voltage peaks and discharges into the load when that voltage falls. If the capacitance is large enough the reduction in charge and capacitor voltage is small before the next peak arrives to top it up. Thus the d.c. level is maintained. The rise and fall in voltage produces a ripple on the supply line at the mains frequency for half-wave rectification and twice the mains frequency for full-wave.

Amplitude of the ripple depends on the load as the discharge is greater with higher loads. Increasing the capacitance reduces the ripple, but it also increases the charging current through the rectifiers on peaks, as they top up the capacitor in a shorter time. The charging current on switch-on is also greater. Higher capacitance thus needs higher rated rectifiers and the provision of small-value series resistors as surge limiters.

The ripple current, which is equal to the total load current, is flowing in and out of the capacitor continuously. It must therefore be designed to withstand this current which is why reservoir capacitors are physically large. Any replacement must be chosen with a ripple current rating higher than the load current at full power.

Half-wave rectifier   The simplest power circuit, it consists of a single rectifier in series with the transformer secondary and the load, with the reservoir across the load. D.c. output voltage is virtually that of the a.c. peak which charges the reservoir, so it is 1.4 times the r.m.s. rating of the secondary. The d.c. current is drawn during the whole of the suppressed negative half-cycle so a much larger current must be supplied by the transformer during topping up. Available d.c. is 0.28 times the a.c. current. The ripple is large, requiring a large reservoir, and more smoothing.

Full-wave rectifier   A centre-tap on the transformer secondary is usually connected to earth. Two rectifiers are wired to the ends of the transformer winding and their cathodes (for a positive supply) are both connected to the supply line and the reservoir capacitor. Only half of the secondary is conducting at one time, so the d.c. voltage is half that of the half-wave circuit, or 0.7 the total r.m.s. a.c. voltage across the winding. D.c. current is equal to the a.c.

Bridge rectifier   Four rectifiers are connected in a bridge-type circuit, with the transformer secondary taken to the anode/cathode junctions of both pairs, while the d.c. supply line negative goes to the two free anodes, and the positive to the free cathodes. It is not always easy to remember which way the diodes are connected in a bridge circuit, so a mnemonic is to remember the *n* in *n*egative and a*n*ode. Two anodes go to the negative, therefore two cathodes must go to the positive; the junctions of each then go to each end of the transformer winding. An alternative is to consider the appearance of the diode, some have a pointed end, others have a line around them at one end. So remember: *lines and points are positive*.

The reservoir capacitor is connected across the positive and negative of the supply. The whole transformer winding is used each half-cycle, so the d.c. voltage is 1.4 times the r.m.s. a.c rating. Current is 0.6 of that taken from the transformer. So for the small cost of two extra diodes, the transformer winding does not need a centre-tap, and the number of secondary turns are halved, although the wire gauge must be thicker to supply the extra current. There is thus a net saving, so the bridge rectifier has virtually taken the place of the full-wave circuit.

Split polarity full wave   A separate positive and negative supply line with a common earth is frequently required for many op amps and output stages. This is achieved with what looks like a bridge circuit but is actually a double full-wave circuit. Four diodes are connected in the same way as for a bridge across the transformer secondary, but the winding has a centre-tap which is earthed. The diodes are thus two pairs of full-wave rectifiers connected for opposite polarity. Voltage for each polarity circuit is the same as for a normal full-wave circuit, 0.7 of the a.c. supply. Current is theoretically half the a.c. supply for each circuit, but if the load is mainly a class B output stage, one section will be supplying high current when the other is passing little. The current rating requirement can thus be considerably reduced.

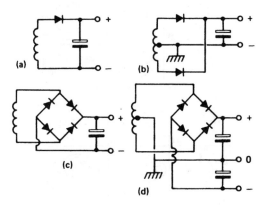

Figure 118. Power supplies: (a) half-wave; (b) full-wave; (c) bridge; (d) split polarity.

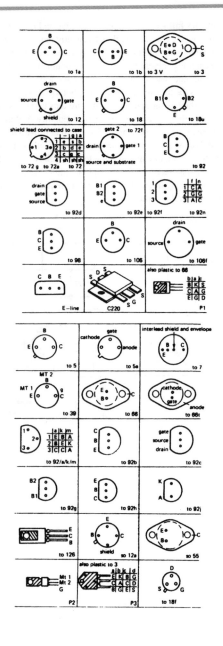

Index